VETERANS
IN STEAM

VETERANS
IN STEAM

COLIN GARRATT

To Stephanie

All the photographs in this book were taken with a Praktica camera, using Pentacon Lenses and Agfa-Colour CT18 reversal film.

The author acknowledges assistance from
World Steam Magazine

First published in Great Britain in 1979 by Blandford Press
Reprinted by Bounty Books in 2005

This edition published in 2014 by Bounty Books,
a division of Octopus Publishing Group Ltd
Endeavour House,
189 Shaftesbury Avenue,
London WC2H 8JY
www.octopusbooks.co.uk

An Hachette UK Company
www.hachette.co.uk

ISBN: 978-0-753728-67-3

A CIP catalogue record for this book is available from the British Library

Printed and bound in China

THIS VOLUME CONTAINS A SELECTION OF LOCOMOTIVES FROM
THE FOLLOWING COUNTRIES

Country	Railway Owner	Gauges
Great Britain	National Coal Board (N.C.B.)	4′ 8½″
German	German State Railways (D.R.)	
Democratic	(Deutsche Reichsbahn)	Metre
Republic		4′ 8½″
Austria	Kamig A.G.	600 mm.
Yugoslavia	Yugoslav Railways (JŽ)	4′ 8½″
	(Jugoslovenske Zeleznke)	
Greece	Greek State Railways (S.E.K.)	4′ 8½′
	(Hellenic State Railways)	
Turkey	Turkish State Railways (T.C.D.D.)	4′ 8½″
	(Turkiye Cumhuziyeti)	
	(Devlet Demizyollazi)	
	Karabük Steelworks	4′ 8½″
	E.K.I. Collieries Catalagzi	Metre
Syria	Syrian Railways	1.05 metre
	(Chemin de Fer Damas Sergayah)	
	(C.D.S.)	
	(Chemin de Fer Hedjaz) (G.F.H.)	

Contents

1 The Ghosts of Blaenavon 9

2 The Hunslet Austerity—A Classic Saddle Tank 23

3 Britain's First 2–10–0s 33

4 Berlin–Dresden: The World's Last High
Speed Expresses 41

5 A Glimpse of German Narrow Gauge 55

6 The Little Fireless of Schwertberg 61

7 A Graveyard in Greece 69

8 By the Lineside at Izmir 77

9 Steam on the Pilgrim Route to Mecca 93

10 A Cottage by the Shed 117

11 Karabük Steel Works 128

12 Along the Black Sea Coast 141

Index 158

Opposite: An Austrian locomotive on Yugoslavian territory. She is of the J.Z. class 53 2-8-2T (Austrian 93 class). First introduced in 1927, Yugoslavia received twenty-nine of these after World War II. This one is caught running in after an overhaul at Maribor Workshops.

1

The Ghosts of Blaenavon

Turner painted the 'Fighting Temeraire' in 1839. Apart from its technical brilliance, the work has a deeply inspiring theme. It depicts one of Nelson's fleet from the Battle of Trafalgar in 1805, being towed across the Thames at Rotherhithe by a brand new steam tug which jubilantly belches a pall of smoke into an evening sky as she draws the condemned sailing ship—pallid, majestic and shorn of all adornments—to its final resting place for breaking up. A blood red sun sets in symbolism on this British tragedy. The age of sail fades against an iron-clad leviathan. Unlike Ruskin and Wordsworth, Turner greatly welcomed the industrial revolution—especially the railway with its fast travel—and he was as content painting glowing furnaces in an iron foundry as undertaking pastoral scenes bathed in opaque mysticism.

Turner's theme for this painting was not simulated; he actually witnessed the scene in 1838 whilst journeying along the Thames on his way to Greenwich. At Rotherhithe, Turner's boat passed the tug hauling the stricken Temeraire and a companion said, 'there's a good theme for you Turner'. No reply was made to this suggestion, but one year later the 'Fighting Temeraire' was exhibited at the Royal Academy. No one would suggest Turner painted exactly what he saw; but he did in essence. The picture's conception and mystique are surely the crystallization of Turner's emotions when confronted with so dramatic a theme which, in all its elements, was close to his heart and enabled him to manifest his patriotism with an unparalleled eloquence. Turner cherished the painting and refused to sell it; upon his death in 1851 it passed—in accordance with his will—to the British nation and it remains in London's National Gallery to this day. Far better than our physically preserving the Temeraire in dry dock, Turner has immortalized her on this priceless canvas.

Turner's painting symbolizes a great historical past; it breathes the British spirit of exploration and simultaneously celebrates the

PLATE 1. *Following pages:* A fine drama at Blaenavon—a once great iron town of the Industrial Revolution. Featured are two Andrew Barclay 0-4-0STs which once belonged to the old Blaenavon Iron Company. Currently in N.C.B. ownership they are 'Toto' (*left*) and 'Nora' built at Kilmarnock in 1919 and 1920 respectively.

industrial revolution: it is a work for historian and romantic alike. I have attempted to transmute some brilliance from this great oil into modern photographic emulsions and, in so doing, evoke another romantic testimony to great deeds of the past.

'Ladies of Blaenavon', which is depicted in the first Plate, is a tribute to one of the world's first iron towns which cradled the industrial revolution. Once the skies above those historical valleys glowed red from countless foundries and Blaenavon's coal and iron provided lifeblood to that revolution in those epic days when British enterprise was second to none. They say the world is girdled with Blaenavon iron: it is almost true. In 1804, the world's first steam locomotive was born a few miles away over the mountains. Today, the foundries and collieries are a memory: derelict and blackened hillsides characterize the area. Blaenavon is a ghost town but a brooding awareness of its florid past bears heavily into the present.

One pathetically small open cast mine survives, worked by the last two steam engines in the area; both are remnants from the old Blaenavon Company. Named 'Nora' and 'Toto' respectively, these little saddle tanks are all that remain of this once prolifically industrialized region. Occasionally these Trojans colour the skies with fire and in so doing recall memories of those halcyon days almost two centuries ago.

Blaenavon lies at the head of the Eastern Valley. Between 1750 and 1800, the South Wales coalfield was developed and became dotted with iron works. The Welsh iron-field was eighteen miles long and one mile wide; it stretched from Blaenavon to Hirwaun. The hills around Blaenavon contained huge deposits of coal, iron ore and limestone; it was an ideal situation for a foundry and in 1789 the first was put into operation. In 1800, over one thousand tons of iron were sent to Newport by canal. Rows of tiny two roomed cottages, built back to back, provided mean habitation for the workers. Little light or air penetrated these buildings and they were so low that occupants had to bend almost double. Large families dwelt amid such hovels.

The misery and exploitation in those hard times is eloquently told in Alec Cordell's novel *Rape of the Fair Country*. The foundries and mines contained men, women and children; women worked underground stripped to the waist; children nine years old descended the pit with their fathers and young boys were left to look after important pieces of machinery—often with disastrous results. Against a backcloth of vile and unnecessary accidents, combined with long working hours in terrible conditions for poor pay, was the Truck System which compelled workers to buy all their food from the company shop. In Blaenavon, this was situated on North Street next to the infamous

PLATE 2. *Above:* In a bluish grey dawn one of Blaenavon's last two steam engines prepares to work the opencast mine. Named 'Toto' she is a remnant from the old Blaenavon Company and was built by Andrew Barclay at Kilmarnock in 1919.

PLATE 3. *Following pages:* The thrilling process of tipping molten waste from the steel furnaces on to slag banks. Performing the duty is a Newcastle built 0-6-0ST of classic design from Robert Stephenson & Hawthorn. She is seen at the Karabük Steel Works in Turkey.

'Drum and Monkey' pub—where many Anglo-Welsh clashes and disputes occurred. The company's prices were invariably twenty per cent. higher than normal and the value of goods purchased was deducted from monthly pay: often when men went to collect their wages they had no money to draw. Strikes and uprisings proliferated and troops were called in to quell disturbances.

The Blaenavon Coal and Iron Company was formed in 1836 and by 1840 five furnaces were in operation producing four hundred tons of cast a week. The railway age had dawned, and Blaenavon's production concentrated upon iron rails; these became world famous and were in high demand. By arrangement with the company, three other British foundries were permitted to stamp their rails 'Blaenavon Iron Co'; this assured immediate sales abroad. In 1857, the famous Crumlin Viaduct was built entirely from the company's iron. Their coal was perfect for locomotives and steamships because of its low sulphur content and was used in large quantities on the world's railways and oceans.

By 1850, Blaenavon had become a pioneer town for trade unionism, but any worker known to be a unionist was sacked by his employer. The company erected a four storey building behind King Street to house Redcoats in readiness for quelling disturbances; rule of rifle restrained even the boldest agitator.

13

In 1879, the works turned to steel production and the company was modified to meet the cost; it became known as the Blaenavon Company Ltd and by 1880 was amongst the finest works in the world. Production thrived with an abundance of collieries; seven blast furnaces and three rolling mills. Thenceforth was a golden age and I quote from a *Short History of Blaenavon*: 'the roar of the works was no discordant noise; it was associated with prosperity. The happy clamour of ringing anvils; the piercing buzz of giant saws; the fascinating rhythms of whirring wheels; the clang and clatter of rolling trams and trucks; the shrill whistle of chugging locomotives; and the sad monotonous call of time telling hooters: all combined to build up a grand awe inspiring symphony of industry. There was nothing like it except the Horeb Choir singing the Hallelujah Chorus. The blackness of night was stabbed at regular intervals as molten furnace waste was tipped, spilling like lava from Vesuvius and painting the clouds to a fiery redness'.

By the twentieth century, Blaenavon was characterized by no less than fifty-two pubs; vicious degradation and self righteous esteem dwelt side by side. In 1900, a strike developed into a long drawn out struggle; families became destitute and many people left town. By 1914 some inefficiencies in management led the company to a decline and their troubles were compounded by Blaenavon having become a national stronghold of trade unionism. World War One provided a temporary boom and Mr Clements—one of the best steel men of his day—arrived as the company's general manager but his endeavours were constantly harassed by union unrest. Many Welsh iron works were closing down as the cost involved in conveying imported iron ore from the coast to the old works up in the valleys became prohibitive once local reserves were depleted; new units next to the sea board were needed. After a sad decline, Blaenavon extinguished its last blast furnace in 1938 and heralded the move towards large units on the coast; the age of Llanwern and Margam was dawning.

Open cast mining continued prolifically for many years, but now only scarred land remains and much of this has been reinstated for new development; in common with many other valley towns, Blaenavon has attempted to attract new industries. Today little evidence exists that Blaenavon's mines yielded millions of tons of excellent coal.

I reached this industrial shrine on a bleak winter's day. A light snowfall had transformed the barren grey wasteland near Forgeside and as I looked across the opencast workings, a plume of steam—pure against a leaden grey sky—indicated a steam locomotive. It was 'Nora' one of Blaenavon's saddle tanks; at least a dozen once worked here and most were named after girls. Reaching the shed, I found 'Toto' and, partly dismantled amid the gloomy depths, lay 'Nan'. All were 0-4-0STs—built in 1920, 1919 and 1916 respectively—by Andrew Barclay in Kilmarnock. Their names were carried on huge brass plates bearing the Blaenavon Co Ltd's name. The 'Co Ltd' has been methodically chipped off – possibly amid Welsh Socialist fervour when the collieries were nationalized and passed into the National Coal Board in 1947!

The shed building was exposed on an otherwise barren waste and the wind howled mournfully through the structures. Doors creaked and whirls of cold air swept in through smashed windows and rattling corrugated iron joints. Every few minutes the roof seemed to give a sickening lurch; here was a place belonging to the past. The foreman was a full blooded Welshman named Garth Williams; Garth is one of Blaenavon's many personalities and has some colourful escapades to his credit; he embodies the spirit of his forefathers as expressed in Alec Cordell's novel. But Garth knew little about the engine's history and bade me contact J. T. Morgan, a local historian with a wide knowledge of the Old Blaenavon Company. 'Make the most of it while you're here', Garth said grimly; 'this lot's coming down soon—they're trying to develop the site—and after a while we won't be using the engines.'

As I left, snowflakes carried on the wind and 'Nora' passed me en route to the shed—her day's work done—I wondered if the person after whom she was named might still be living in Blaenavon. Who were 'Nora' and 'Nan'? It needed little effort to imagine the rigours and misery of working here centuries ago. The wind blew as cold as charity over the hillsides and so grey was the afternoon that twilight had begun by three o'clock. With my coat well buttoned, I struggled down the hillside towards town, passing on the way an abandoned steam crane; its derelict form looking like the skeleton of a dinosaur. I reached Broad Street half frozen and arrived at the 'Lion' where a roaring coal fire in the snug quickly restored me. This hotel was attacked in 1868 by what Blaenavon's history describes as a boozey maelstrom who, knowing mine host to be an active member of the Tory Party, wrecked his premises and sent barrels of beer, wines, spirits, and cider rolling and gurgling away down Broad Street.

At the Lion, I met J. T. Morgan—a sprightly ninety year old. And what a tale he could tell; 'we were the workshop of the world when I was a boy' he said. He described the fiery skies and the coals being exported to Cardiff Docks for locomotives and steamships throughout the British Empire; recalled the huge train loads of iron and steel leaving Blaenavon bound for Pontypool; and took pride in telling me how the tyres for the

PLATE 4. *Following pages:* In this typically industrialized setting at Pennyvenie Colliery West Ayrshire we find a standard Andrew Barclay 16 in. 0-4-0ST undertaking her workaday chores. Built as recently as 1951, she is seen leaving with a loaded train bound for the B.R. exchange at Waterside.

17

G.W.R. 'King Class' were made on Forgeside. Yes he remembered all the Ladies of Blaenavon: 'Betty', 'Chrissy', 'Gina', 'Jean', 'Joan', 'Lily', 'Nan', 'Nora' and another engine named 'Toto'— all saddle tanks which came from Andrew Barclay of Kilmarnock because Mr Clements and his engineer were Scottish. He could not tell me who the girls were but promised to find out.

THE BLAENAVON NORA. Nº.5.

BLAENAVON NAN. Nº.18.

At nine o'clock the following morning, J. T. Morgan called at the 'Lion' to escort me over the surrounding hills to visit the old coal, iron and limestone mines. The weather remained bitter, but my guide, well muffled up, strode on unconcernedly with the vigour of a man half his years. He had been speaking with Miss Lily Jones—once secretary to Mr Clements—and she claimed one engine was named after her, whilst Nora was daughter to another manager. She also suggested that engines were named after Mr Clement's four daughters and that Toto had been their family dog. J. T. Morgan said he would verify this with his friend Mr E. Morgan and let me know by letter.

This forgotten town is steeped in its origins; it reeks of passion and drama. The furnaces are ruined, the labourers' hovels have gone; yet spirits from the past cling defiantly to those grey wasted hillsides. Alec Cordell felt them too and though his novel was set in the 1820s, it has an intrinsic meaning in Blaenavon today. I began to smell sulphur in the wind. Was that the dell where scruffy Sarah Roberts sat chipping rock from iron veins since she was five? And that golden heather stretching away towards the Usk Valley; is Morfydd up there with some man, winter or not? Her sensuality satisfied, but her burning hatred for the ironmasters running hotter than the furnace which poured molten iron over nine year old Enid Griffiths' legs.

The half frozen slag crackled beneath my feet as I walked the hillsides. Is this the cave where little Ceiwen Hughes died of cold and deprivation after breaking coal from dawn to dusk through that terrible winter? Her mother caught the iron on her shoulder a year previously and lay paralyzed ever since; Ceiwen's fourpence a week helped the stricken family; here is where they found her—a scruffy ball of rags and flesh.

I passed a water butt standing near the old forge and thought of Morfydd's brother Iestyn on his first day at the foundry when he was eight. He rose long before dawn; the valley was baked in a hard frost and whilst Morfydd vitriolically attacked his father for sacrificing him to the ironmasters, Iestyn passed outside to wash from the butt. The novel reads—'he plunged his body into the frozen water; rime was lipping the tub. To know the shock of lost breath and fight to get it back. Trickles of freeze run down your blue chest and soak the waist towel; splash and thresh about and take great breaths of white mountain mist. Down into the lungs it goes making the blood run in hot agony; rub, rub with a towel and sing for courage. No hair on the chest or belly like Dada, but it will come after a month at the foundry where grown men die in the heat and frost says Morfydd. "Bore da chwi", shouts Twn-y-Beddau, the coal trimmer from next door. Naked as a baby he is and his children throwing buckets at him. "Good morning" I shout back shivering, "up at the top today Iestyn?" "aye, aye".'

Continuing, I discovered the thickets where Iestyn's father was beaten and whipped; his shirt flogged into bloody shreds over his trews and his house ransacked because he would not join the union. The Chartists were becoming established; better times were being forged but few of the rugged stalwarts would live to see the raw iron of their dreams built into an enduring monument.

Looking back across Blaenavon with its cold grey roofs and wet streets, I reached the place where J. T. Morgan had brought me the previous day. Iestyn and Mari Dirion were in this meadow when the storm drove them into Shant-y-Brain's hayloft to that strange and wonderful first loving.

Their blood ran hot with the kiss, their hearts hammered with quickening beats; cattle lowered over the farm; rain swished across the heather and, under a flickering crimson cloud, came the distant thuds from Blaenavon forge. 'Oh', whispered she, 'there's wicked I am; and you for doing it—this time next year I will be walking twins around town.' 'They will beat you up the hills with sticks and bibles months before that', Iestyn said kissing her.

I received a letter from J. T. Morgan shortly after my return home. He had made many enquiries in the valley but some names seemed irretrievably lost.

2

The Hunslet Austerity - A Classic Saddle Tank

Not another chapter on Hunslet Austerities you say! A reasonable comment I admit; yet classic stories, by their very nature, will withstand a repeated performance; albeit with slightly different emphasis and interpretation. I will defend myself further by stating this class to be one of the most popular; are there not more Hunslet Austerities preserved than any other class in the world? And, apart from their fascinating history, the type can still be found in commercial service today.

Having pleaded some justification, the Austerities origin goes back to the days when Europe was war-torn. Britain was united in fear of the Nazis and all hope was placed on the long awaited invasion of Western Europe when the oppressors might be driven back. New locomotives were urgently needed for this campaign; a main line goods engine and a heavy shunter. In 1942, a conference was called at which Edgar Alcock of Hunslet met the Minister of Supply and R. A. Riddles (deputy director of Royal Engineers equipment). He convinced them that a war-time design based upon Hunslet's modern saddle-tanks would be superior to the L.M.S. Jinty for shunting purposes.

Alcock won the day, and in view of the urgency, it was decided that Hunslet were to cease producing armaments and begin work on the saddle-tank immediately—over the war years, locomotive workshops were extensively involved in building tanks and munitions; normal locomotive maintenance having to take a secondary rôle. The first Austerity saddle-tank was steamed at Hunslet's works on the 1 January 1943; only four and a half months after the conference.

Hunslet's engine was accepted for two principal reasons: its simplicity of design; and a shorter wheelbase for greater availability. Considerable economy in materials rendered the new engines as true Austerities. Though based on an earlier Hunslet design, the driving wheel diameter was increased to 4 ft 3 in. for short journey military trains and an increase in coal and

water capacity was provided. Superb engines, they were easily capable of starting 1,100 ton trains on level track and 550 ton trains on a 1 in 100 gradient. Principally allocated to military establishments and docks, others went into key industries such as collieries. Between 1943/7, three hundred and seventy-seven of them were produced for the War Department from six different builders.

After the invasion in June 1944, many went abroad and did service in France, Belgium and Holland. Not all returned to Britain: twenty-seven were retained by the Netherlands Spoorweg; some of these passed to the Dutch State Mines and survived in Holland until the 1960s. Others remained in France.

By 1946, the W.D. began to dispose of surplus engines. Seventy-five were incorporated into L.N.E.R. stock—forty being new locomotives stored at Longmoor. Classified J94, they were numbered 68006–68080 under British Railways and were found at a variety of sheds, notably Immingham, York, Darlington, Sunderland and Bidston in North Wales. Many passed into Colliery ownership—especially those allocated during the war. After the National Coal Board was formed on 1 January 1947, further engines passed to the coalfields including many returned from overseas. Others passed to steelworks and docks and six were sold to Tunisia in 1946. By 1952, less than one hundred remained under W.D. control although fourteen new ones were built in 1952/3. Their duties centred around military establishments such as Longmoor and Bicester.

As we have seen, the National Coal Board were fortuitously provided with a modern standard type to replace the multitude of ageing veterans inherited from private companies. Having acquired as many W.D. engines as possible, the N.C.B. ordered new ones and building continued until 1964, by which time a grand total of 484 engines had been reached.

It is remarkable that a locomotive, designed for war-time service, and not intended for long term operation, should be built over a twenty-two year period and—even more remarkable— that it should outlive all main line classes by some fifteen years! Nevertheless, the ex L.N.E.R. J94s were broken up from 1960 onwards, and by 1964 when the last Austerity was built, they had almost disappeared; only six passing into industrial service. I cannot think of any other class in world history that was being built and withdrawn simultaneously. Shortly after 1960, I remember seeing some dumped at Woodham's Scrapyard in Barry. Decked in blue, they had come from the Longmoor Military Railway and had L.M.R. boldly stamped across their saddle tanks: which to the uninitiated, meant London Midland Region.

PLATE 6. *Opposite:* Another rural scene featuring a Hunslet Austerity—this time in Scotland. One of the N.C.B.'s Giesl-Chimneyed variants is seen hauling a loaded train from the B.R. exchange to the Rexco Plant near Saline.

Most railway enthusiasts are interested in the variety of liveries, lining out styles and letterings, variously applied to classes throughout their working life. But the intrigue is largely retrospective, since any particular style currently in vogue creates an illusion of permanence; especially when successive engines are seen to be similarly treated. Variety is greatest when one class has many different owners as with industrial engines— but few have been so prolifically varied as the Hunslet Austerity. Throughout their many phases of ownership, countless styles have occurred over a thirty year period in addition to a plethora of livery changes and renumberings on account of transfers. Many have been named at various points in their career especially those in W.D. service; fine names such as: Sapper, Greensleeves, Arnhem, Royal Pioneer and Lisieux, to mention a few. Their rectangular nameplates were superbly cast complete with a badge and were carried on the engine's running plate above the centre driving wheel. However, my personal favourites were those named in the best of L.M.S. 'Jubilee' traditions by the North Western division of the N.C.B.: Warspite, Revenge, Warrior and Hurricane—suitable appendages for these pugnacious rearguards of British Steam. These nameplates were also rectangular, but affixed to the saddle-tank, and the engines were finished in an attractive maroon livery.

Various structural modifications provide a further dimension for study; Westinghouse brake pumps and cylinders on many W.D. engines, Giesl Chimneys for reduced coal consumption, Mechanical Stokers and Steam Generators for electric light. Perhaps the greatest modification was Hunslet's 'Special Blast Pipe' fitted to reduce the emission of black smoke. This involved considerable internal rebuilding and was undertaken by Hunslet for the N.C.B. during the early '60s—a time when public outcry against smoke and pollution was rampant.

Such diversity within one type makes the study of steam locomotives deeply rewarding; the variations from one to another are infinite; no two are ever the same. Compare this with species in the animal kingdom; certainly variations often occur during the lifespan—possibly differences between immature male and female—but these are absolute and often limited in number with most individuals appearing identical.

A friend of mine has ambition to document the Austerities' history in its entirety—setting out the life and fate of each engine. He is also a fine watercolourist and one imagines the evocative pictures he would paint. Should his ambition ever be fulfilled he will have created a classic work.

During the summer of 1977, over forty Austerities remained in commercial service; many were standby engines to cover diesel breakdowns and overhaul periods, though some were active on a regular daily basis. All came under the N.C.B. auspices except three with the Ministry of Defence. None survived working abroad. Their industrial distribution today follows that of the prime years and may be generalized as South Wales, Derbyshire, Lancashire, Yorkshire, Northumberland and Durham. They have occurred less prolifically in Scotland although two collieries there still employ Austerities in regular operation.

To children in our schools today, the steam age is a part of deep history: certainly no child under thirteen can remember seeing a mainliner and yet some forty Hunslet Austerities remain on the active list throughout Britain. More remarkable is their preservation; over thirty have been saved and, as more examples are purchased from the Coal Board's roster, their lead in preservation is never likely to be surpassed. Perhaps the best place to see one is on the Keighley and Worth Valley Railway; here a real W.D. example can be found at work. She is 'Brussels'; painted in W.D. blue complete with authentic nameplate, Westinghouse brake equipment, steam generator and L.M.R. blazoned on her saddle tanks. She is a splendid working exhibit and is vigorously put through her paces over the Worth Valley's 1 in 58 bank.

A few words on some of the colour plates: No. 5 celebrates the closing years of steam on the Backworth system north of the River Tyne—a place especially noted for Austerities over post war years. It will be remembered that William Hedley put some of the world's first steam engines on to this historic coalfield in 1813; almost $1\frac{3}{4}$ centuries later, the Hunslet Austerity was destined to bring that great tradition of steam and coal to a close in the north east. Backworth was steam's last stronghold in Northumberland and, well into the 1970s, Austerities could be found working round the clock taking coal from Eccles and Fenwick pits to the exchange sidings with British Rail. This sunrise picture depicts ex W.D. No. 5161—built by Bagnalls of Stafford in 1944 leaving Fenwick colliery on a December day with the first haul of the morning shift. She is running as the N.C.B.'s Northumberland Area No. 6.

Plate 6 set in rural Fifeshire, shows one fitted with an elongated Giesl Ejector. This engine's identity is something of a mystery. It was one of those unfortunate occasions when I omitted to take proper notes. A good clue is provided by the ornately painted No. 5 on her buffer beam, but lists of Scottish locomotives in my possession mention no such engine. Several Austerities working at English collieries during the early '70s were numbered '5'— each within their respective divisions—and possibly one of these was transferred to Saline but, as I am unfamiliar with the lettering

PLATE 7. *Following pages:* Golden January lighting invigorates the Northumberland landscape as a Hunslet Austerity 0-6-0ST heads back to Shilbottle Colliery with a rake of empties from the B.R. exchange near Alnwick. A magpie sitting on the foreground fence would have completed the scene.

styles, nothing certain can be gleaned. The moral is always carry a notebook and pencil when exploring locomotives; one invariably needs to refer back in years to come. Suffice it to say she is an extremely handsome engine.

One Austerity which needs no identification is the subject of Plate 7. She is ex W.D. No. 5077—built by R.H.S. of Newcastle in 1943—and running here as N.C.B. Northumberland Area No. 45 at Shilbottle Colliery. Although ex W.D., neither this one, or the Fenwick engine, ventured abroad during war service; and neither were ex L.N.E.R. J94s. Thus they passed directly to the Coal Board from the W.D.

Shilbottle is situated north of the Tyne near Alnwick. Set amongst some of England's most beautiful countryside, this colliery was one of the last to dispense with steam traction in the north east. The picture with its foreground dominated by yellow winter grasses, is bathed in the golden lighting of an English January afternoon. I allowed the engine to extend well across the scene to evoke a feeling of movement, whilst the two waggons were counterbalanced by two trees on the left; I was pleased with the result.

But when I brought it home and showed it to a friend she said, 'Yes it's all right.' 'All right', I echoed, continuing to extol its virtues, 'surely it's more than all right?' 'Well', she continued, 'there's something missing.' Again I looked at the picture; it seemed complete to me. 'What's the matter with it?' I asked; and she told me, 'There must be a Magpie sitting on the foreground fence; and if the bird were looking across to the train your picture would be a masterpiece'; but that without the bird she did not think it so. She was right; but little did she know that two Magpies were flying around that coppice during the hour I stood waiting for the train.

HUNSLET AUSTERITY 0-6-0ST

Building Sequence	Builders Involved	Leading Dimensions
Built 1943–47 for W.D. 377	Hunslet, Leeds	Cylinders 18 in. × 26 in.
Built 1952–53 for W.D. 14	Hudswell Clarke, Leeds	Boiler Pressure 170 lbs sq. in.
Built 1945–64 for collieries and steelworks 93	Vulcan Foundry, Lancs	Grate Area 16·8 sq. ft.
———	Yorkshire Engine Co, Sheffield	Axle Loading 16·35 tons
Total: 484 ———	Robert Stephenson & Hawthorne Newcastle-on-Tyne	Tractive Effort 23,870 lbs
	W. G. Bagnall, Stafford	
	Andrew Barclay, Kilmarnock	

3   Britain's first 2-10-0s

During World War Two, enormous pressures were imposed upon Britain's railways; the vast movement of wartime supplies both at home and in Allied countries, meant an inevitable shortage of locomotives. A European invasion was also planned and suitable motive power would be needed to follow the advancing armies. Over the initial war years, Stanier's 8F 3-8-0 was adopted as Britain's official war engine and the ranks of this already well established class were greatly swelled. Satisfactory as the 8F was, an austerity version was decided upon; an engine suited to immediate production—with an emphasis upon saving as many man hours and scarce materials as possible.

The first Austerity 2-8-0 appeared in 1942 and was joined that year by America's counterpart—Major Marsh's famous S160s—which although intended for wartime operation on the widest scale had been specially built to the British loading gauge. The Austerity 8F was designed by the North British of Glasgow to specifications laid down by R. A. Riddles—then deputy director of Royal Engineers Equipment—and, by the close of 1943, four hundred and fifty were in service having been built solely by the North British and Vulcan Foundry, Lancashire. Additionally some four hundred S160s were also in traffic.

Having provided a basic Austerity, Riddles felt that a larger version was necessary; an engine having identical power but a greater route availability. These were to be Britain's first 2-10-0s. They had a larger boiler and wider firebox and proved to be very free steaming indeed; their middle pair of wheels was flangeless and the axle-loading two tons less than their 2-8-0 counterpart. Both types of engine had very much in common, including identical tenders.

One hundred and fifty 2-10-0s were built; all by the North British between 1943/5 and the last two were named after this company. Pending the great invasion of Europe, some were 'run in' by the L.N.E.R. and L.M.S., particularly the latter, and they became familiar performers along the ex L.N.W. and M.R. main lines working from such sheds as Willesden, Wellingborough and Toton.

Inevitably these War Department engines aroused much interest in the country; Britain had finally aspired to a 2-10-0 no less than twenty-five years after the type was introduced in Germany with the Prussian G12 during World War One. Indeed, only two ten-coupled designs had ever appeared in Britain before and both were unique engines; Holden's 0-10-0T for the Great Eastern Railway and the famous 'Big Bertha'—the Lickey Banker 0-10-0 introduced in 1919.

Early in 1943, twenty W.D. 2-10-0s were shipped to the Middle East. Four were retained by the Chemin de fer Damas Hamah et Prolonguements railway in Syria, whilst others went to Greece to become the Hellenic State Railway's Lb class taking the numbers 951–966. Shortly afterwards, the European invasion began and the 2-10-0s along with many of their 2-8-0 relations and S160s went abroad. One engine however, the celebrated 'Gordon' stayed in Hampshire on the Longmoor Military Railway.

The great invasion was a success and war service for the Austerities was over by mid-1945; a decision had to be made upon their future distribution. One hundred and three 2-10-0s went to the Netherlands Spoorweg; eventually being purchased by that railway. Surprisingly they had a very short life; all were withdrawn by 1952. This left only twenty-seven of the original hundred and fifty for Britain and after a period of storage on the Longmoor Military Railway a couple were tested over the Scottish region of the L.M.S. As a result, twenty-five were transferred there and taken into British Railways stock in 1949/50: they were numbered 90750–90774. These twenty-five engines made a remarkable contrast with the seven hundred and thirty three W.D. 2-8-0s taken over by British Railways in 1948. A couple of 2-10-0s remained at Longmoor; 'Gordon' and 'Kitchener'—as the other engine was named; they were numbered W.D. 600/1 respectively.

The Longmoor Military Railway in Hampshire was conceived early this century and became the principal base for training military personnel to operate railways in foreign lands. Over two world wars, Longmoor proved a great asset; locomotives and equipment being shipped out directly to strategic areas abroad. A wide range of engines has worked the system and when operations at Longmoor reached their peak during World War Two some twenty-seven engines could be in steam simultaneously. In later years motive power at Longmoor was reduced to standard W.D. classes. Although a self-contained railway, two

PLATE 8. *Opposite:* The first British 2-10-0 was the Riddles W.D. 'Austerity', first introduced in 1943. A few of these historical engines still survive in Greece and one is shown here taking water at Alexandropolis.

connections were made with the Southern; one at Borden and another at Liss on the London–Portsmouth line. Longmoor was closed after Ministry of Defence cuts in 1969; many believed it should be converted into an operational museum line and bold attempts were made by preservation groups. Sadly this did not materialize and today Longmoor is without locomotives or serviceable track.

The only other 2-10-0s to work in Britain were the standard B.R. 9Fs, also designed by Riddles, and introduced in 1954. These fine engines are noted for their remarkably short life; though building continued until 1960, all had disappeared by 1967. However, they fared better than most of their Austerity predecessors, as by 1952, two thirds were condemned—only eight years after the class's inception! Again, the comparison with Germany is remarkable as, forty years after the introduction of the Reichsbahn 50 Class 2-10-0 in 1938, countless examples remain at work today in several different countries.

Fortunately some W.D. 2-10-0s survived in Britain to be enjoyed after the war and, during the early 1950s, the Scottish Region batch were allocated as follows: Motherwell 13, Grangemouth 8, Carstairs 2 and Polmadie 2. They also worked from Carlisle (Kingmoor) depot.

My first acquaintance with the Austerity 2-8-0s was in 1950 when they were regular performers along the Midland main line but, as young boys, we were greatly intrigued to learn that a 2-10-0 version existed in what, to us, was far away Scotland (we had not heard of Longmoor at that time). A few years later I made a visit to Scotland and a special stop was made at Motherwell to see one of these 'fascinating beasts'. I remember arriving well after dark on a damp summer evening; my eagerness would not allow me to wait until morning so I set off for the sheds without further ado. It is amazing how certain events from years past retain perfect clarity in the memory, for it might have been only last evening when I stumbled across that expanse of wasteland towards the shed.

The yard was visible; smokey outlines of engines shimmered under the lamps and, to my joy, I picked out the silhouetted form of the big Austerities; their elongated boilers and stubby cast iron chimneys—so tiny in proportion—being unmistakeable. I had counted four prior to actually reaching the yard; surreptitiously climbing a bank—the last obstacle to my goal—I rapidly noted down their numbers—lest the night foreman should catch me. What a contrast they made with the old tall chimneyed Caley tanks and goods engines. It was a thrilling night for a young boy who had found his first 2-10-0s: little did I dream that it would be almost twenty-five years hence before I saw the last of this type in the Syrian desert. Those Scottish engines were withdrawn and scrapped in 1961/2.

Upon the dissolution of Longmoor, 'Gordon' had become something of a national attraction; decked out in her latter day blue and red livery, she looked an exceedingly handsome locomotive and had been greatly admired at Longmoor on open days. Most people heralded 'Gordon' as the last survivor; few pausing to think of the twenty engines sent to the Middle East in 1943—such was our preoccupation with home railways during the 1960s. But these were the ones destined to have an average life: all remaining in service until the mid-seventies.

The Greek engines were confined to the north and, amongst other duties, worked international services for Istanbul up to Pithion on the Turkish Frontier. Here a remarkable event occurred; they came face to face with their German war-time equivalents. This meeting of British and German war engines continued until 1973, when the 07.48 from Thessalonika, hauled by a Lb, met the 07.13 from Istanbul, worked by a Turkish 56 Class or D.R. 52 'Kriegslokomotiv'. It is fascinating that British and German freight designs for war-time should meet working international passenger trains, thirty years later, in Greece and Turkey respectively! Thessalonika's Lbs also worked up to Gevgeli on the Yugoslav border.

During the spring of 1973, I made a journey to Greece to see the 'Lb' in action, but having reached Thessalonika, had become so ill as to be confined to my hotel bed. After a couple of days it became apparent that my illness was going to last the remainder of the trip; so, on the penultimate evening, I got up, staggered down to reception, and called for a taxi out to a level crossing two miles distant. I knew a 'Lb' would work the evening passenger train to Drama and was determined to see one. It was twilight before the crossing gates finally rattled down and excitedly I scanned the long straight leading to Thessalonika's main station. A blazing headlamp became visible; palls of steam rose up above the buildings and a musical exhaust became audible. She had worked up to a lovely chattering rhythm, as with a deep toned whistle call, she swept over the crossing with a swish of steam and a ringing 'plonk' from the motion—a delightful characteristic of the Austerities and one well remembered by all who knew them. As the lit coaches passed, I marvelled at this British exile; few passengers on board would have imagined their engine to have been part of Britain's war effort in 1943! I watched her arc round a bend and go dashing away through the outer suburbs with steam billowing up into an azure sky; her steady beat eventually becoming drowned by traffic noise as the crossing gates lifted.

By 1976, all Greek Lbs had finished regular passenger duties;

many lay abandoned. Several remained in working order at Thessalonika but were only used on winter snow plough duties. 'Marvellous engines', the depot chief told us looking affectionately at No. 960: 'built to last five years and still here thirty years later', he said. An old fitter claimed to remember them arriving in Thessalonika by boat from Egypt and said they were given to Greece in deference to her cooperation with the Allies.

Fortunately, one was kept as steam standby at Alexandroupolis close to the Turkish border, and after a wonderful six hour journey following the Aegean coast line, we reached our objective: a town much frequented by tourists attracted by the glorious beaches and architectural ruins nearby. The depot yard was alongside the Aegean; two 'Lbs' were present: No. 964 in steam; No. 962 dead. Throughout that summer, No. 964 stood in steam continuously but never turned a wheel; there were sufficient diesels to cover all requirements and she was left, warm and hissing like a phantom from the past. What memories they evoked; both were grimy all over—exactly as they were when they, and their 2-8-0 counterparts, ran in their hundreds all over Britain; I know of no other class in British history which was so consistently dirty.

Was it really twenty-five years since that night at Motherwell? A pair of white storks had built a huge nest on an out-building and the sight of these enormous birds gliding around a shed yard on the Aegean Sea, with a British Austerity and a Mediterranean sunset, was almost surrealistic in atmosphere.

The nostalgia I felt upon leaving Alexandroupolis was mitigated by the knowledge that four were still in Syria. They had been absorbed into the Syrian State Railway in 1948 and were known to be still operating—having become a popular topic of conversation amongst railway enthusiasts. The disappointment we were to encounter upon reaching Syria is enumerated later in this book; suffice it to say here, that No. 964 was the last I would see in steam.

Two have been preserved: the least known being in W.D. livery at the Netherlands Spoorweg Museum, Utrecht. Named 'Longmoor', she was the 1000th British built freight locomotive to be ferried to Europe since D-day. Her preservation in Holland is fortuitous considering the N.S. regarded their Austerities as little more than a stop gap. After Longmoor closed, 'Gordon' passed to the Severn Valley Railway at Bridgenorth and, in contrast with her Dutch sister, remains in blue and red. During the summer of 1977, Gordon was a star attraction at the Severn Valley Railway's enthusiasts' day; eleven engines were in steam and five trains in continuous operation, including one hauled by a Stanier 8F—predecessor of the War Department Austerities.

PLATE 9. *Above:* The world famous shape of the Reichsbahn 'Kriegslokomotive'—class 52. The T.C.D.D. operate some fifty-three of these engines and sometimes employ them on the 09.45 Izmir–Denizli 'mixed'. Here with this well-known train is one of the breed making a fine contrast with the teeming suburbs of Izmir.

Shortly after steam became a memory on B.R., it was realized that no Austerity 2-8-0 had survived; all seven hundred and thirty-three had been broken up between 1959 and the late sixties; it seemed ironic that two of the much less numerous 2-10-0s should have been saved. The Austerity 2-8-0s were largely ignored by enthusiasts who, inevitably turned their attractions upon preserving designs with a better pedigree. But as time passed, the Austerities' absence was felt: they became a mourned class. Almost one thousand had been built—a modern design of considerable significance—yet they were not only grossly under photographed, but extinct! Then came the miraculous discovery of one stored inside Sweden's Arctic Circle; it had originally worked in Holland and, along with a sister engine, had been sold to Sweden as precedent for a bulk purchase which did not materialize. Classified G11 by the Swedish State Railway, she had long been inactive. Negotiations were begun for the engine's return to Britain and she was purchased by four members of the Keighley and Worth Valley Railway over whose metals she works today.

Another miracle occurred in 1967 when arrangements were finalized for an American S160 to be brought to Britain from Poland; this was the missing link in a quartet of great war engines. With imagination and foresight we have transformed these celebrated locomotives into living museum exhibits. Long may they continue to provide historical interest and aesthetical pleasure. The four are enumerated below:

BRITISH AND AMERICAN ENGINES FOR WORLD WAR II COMPARATIVE DIMENSIONS

Class	Wheel Arrangement	Cylinders	Boiler Pressure
Stanier L.M.S. 8F	2-8-0	$18\frac{1}{2}$in. × 28 in.	225 lbs sq. in.
Riddles W.D.	2-8-0	19 in. × 28 in.	225 lbs sq. in.
Riddles W.D.	2-10-0	19 in. × 28 in.	225 lbs sq. in.
Major Marsh S160 U.S.A.T.C.	2-8-0	19 in. × 26 in.	225 lbs sq. in.

Class	Driving Wheel Diameter	Grate Area	Axle Loading	Tractive Effort
Stanier L.M.S. 8F	4 ft. $8\frac{1}{2}$ in.	28·6 sq. ft.	16 tons	32,438
Riddles W.D.	4 ft. $8\frac{1}{2}$ in.	28·6 sq. ft.	15.6 tons	34,215
Riddles W. D.	4 ft. $8\frac{1}{2}$in.	40 sq. ft.	13·3 tons	34,215
Major Marsh S160 U.S.A.T.C.	4 ft. 9 in.	41 sq. ft.	15·6 tons	31,490

4 Berlin-Dresden: The World's Last High Speed Expresses

The year 1844 was a significant one in railway pictorialism; Turner produced 'Rain, Steam and Speed'. This painting was possibly the first to depict a steam train in full cry and it remains today as tribute to the majesty of high speed travel with steam. The Bristol and Exeter extensions of Brunel's seven foot gauge main line completed some two hundred miles of continuous rail travel; speeds of ninety miles per hour had been reported: Turner was enthralled. His picture shows a Gooch 'Firefly' class 2-2-2 speeding over Maidenhead Viaduct in a squally storm. The 'Firefly' sweeps towards the viewer at an alarming rate; flaming coals bounce alongside the engine; three puffs of steam recede ethereally into the distance and crowds wave from the rain sodden river bank—in 1844, railways were still a novelty to the populace at large. The celebrated hare races for its life ahead of the locomotive—the machine's speed outstripping that of nature. Man had never travelled so fast.

Down in distant meadows, horses draw the plough; their sedate movement symbolizing the quiet timelessness of centuries and making ultimate contrast with the dashing 'Firefly' as it bursts through a misty haze in the Thames Valley. The coaches recede to a breathtaking perspective, terminated by closing shrouds of vapour: a miraculous presentment of movement, mist and moisture. On 8 May 1844, *The Times*, commenting upon this painting, said: 'whether Turner's pictures are dazzling unrealities, or realities seized upon at a moment's glance, we leave his detractors and admirers to settle between them.'

At the Royal Academy exhibition that year, Thackeray, upon seeing 'Rain, Steam and Speed' claimed that Turner had 'out prodigied all former prodigies.' He has made a picture 'with real rain behind which is real sunshine and you expect a rainbow every minute. Meanwhile, there comes a train upon you moving at fifty miles an hour [Thackeray's estimated speed] and the observer had best make haste to see it lest the train should dash out of the picture and be away up Charing Cross through the wall opposite.'

PLATE 10. *Following pages:* One of the best loved classes of today are the German 01 Pacifics. They work some of the last high speed steam trains in the world between Berlin and Dresden in the G.D.R. Here is one seen at speed near Bhola with a twelve coach express on a mile a minute timing from Berlin.

In common with Turner's epic 'Steamer in a Snowstorm', 'Rain Steam and Speed' has become a controversial painting, even by Turner's standards. It was bequeathed to Great Britain and has been in the National Gallery since 1865.

For over a century, the fleeting brilliance of high speed steam captured the world's imagination; railways were the premier technology; the steam locomotive hastened social change with unprecedented alacrity. Whether the 90 m.p.h. allegedly achieved on Brunel's seven foot gauge is true or not; it is now generally believed that a three figure speed was reached in 1904 when, amid rivalry between the London and South Western and the Great Western to bring the Ocean mails up from Plymouth to London, the Great Western 4-4-0 'City of Truro' made that historic 100 m.p.h. dash down Wellington bank.

Some quarter of a century later in 1932, Great Western 5006 'Treganna Castle' ran the 77·3 miles from Paddington to Swindon in 56 minutes 47 seconds—an average speed of almost 82 m.p.h.—and maintained 90 m.p.h. for 38 miles on end. The performance achieved by L.N.E.R. A4 'Silver Link' is perhaps even more stunning; twice she attained $112\frac{1}{2}$ m.p.h. on trial runs for the Silver Jubilee in 1935—averaged 100 m.p.h. for 43 miles continuously—and 108 m.p.h. over the $10\frac{1}{2}$ miles between Biggleswade and St Neots. Such capabilities enabled the A4s to obtain the world speed record for steam traction, when on 3 July 1938, A4 'Mallard' ran at 126 m.p.h. with a 240 ton test train on Stoke bank between Grantham and Peterborough.

Scintillating and exceptional as these performances might be, they do show the greyhound like qualities to which a fine steam locomotive and crew can aspire, albeit with special trains, or even to mark steam's passing—as in 1967—when Bullied's ex S.R. Pacifics were whipped up to three figure speeds in a final triumphant blaze of glory and enthusiasm.

The steam train offered much to the artist; form, colour, fleeting movement, fire, smoke, and steam; all set in tension with the elements against an ever changing backdrop for every quarter mile run. But few painters turned to the railway for motifs. Certainly there was a predominant attitude during the nineteenth century that industrial subjects were not suitable for art and in May 1850 the pre-Raphaelite journal made a plea for more technological motifs; but to no avail.

Although some topographical views of trains were delightfully executed, notably J. C. Bourne's, the subject has been largely ignored by the more imaginative painters with the exception of Claude Monet. Indeed, depiction of steam trains was to become almost entirely the photographer's preserve; especially during the twentieth century, when it is no exaggeration to say that railway photography has been one of the most popular—and ever recurring—subjects in the medium. It is fitting that this long pictorial tradition in photography should be precedented by a great painting. Only recently have painters taken up the railway theme with any relish, but the rash of popular prints subsequently produced have—perhaps inevitably—been somewhat photographically orientated and they imbibe few of the imaginative and spiritual qualities inherent in great painting.

The visual glorification of steam railways is hardly surprising; few people would not be arrested, albeit momentarily, by a powerful locomotive speeding by with a heavy express. Over the golden years of train watching, it was the expresses which captured the imagination; an L.M.S. Duchess thundering through Nuneaton with a seventeen coach express—newspapers lifting from bookstalls and swirling away up the platform in its wake; L.N.E.R. A4s running at almost 100 m.p.h. through the Soke of Peterborough, their chime whistles drifting across the lush countryside; a G.W.R. King dashing along the Vale of White Horse on its way westwards with sunlight glinting from brass mountings and nameplate; or air-smoothed Bullied Pacifics, howling their way through Basingstoke with boat trains for Southampton.

These are but a few infinitesimal memories from a golden era which thrilled and inspired us; we could never imagine life without steam trains. And why should we? They were part of our heritage; and by logical process would be with us for ever.

For millions, train spotting—regarded as futile only by boorish intellectuals—offered colour, romance and excitement; its greatness far surpassed superficial appearances. With over 20,000 steam engines once active in Britain, covering hundreds of different types, it offered sport indeed; it opened young eyes to the geography of one's country and gave incentive to travel; it revealed industrial history; it developed the eye for detail and aesthetics; the mind for numbers and statistics; and it broadened comprehension of distribution and purpose; cause and effect. He who understood railways had his finger on the nation's pulse. The benefits from so fine an educative process remain for life and one of the many advantages I have personally inherited is an ability to remember telephone numbers—blessed and unforgettable is he with number 60022! It is pointless to continue, I am understood well enough.

In the 1950s, every other teenager was a loco-spotter; during evenings, weekends and holidays we would flock to the linesides; there was always companionship at one's local bridge or cutting. During summertime, the famous spotting places on main lines would attract hundreds of people; grass was worn off the embankments and we sat on hard dusty patches. Cycles, popbottles, sandwiches, notebooks and A.B.C.s were the vital elements. The railway banks were a grandstand to one of the finest unfolding dramas of all time and as in great sport, the thrill of the unexpected loomed behind every quiet moment; thirty classes might be seen on one day, and from these any rarity could pass. I remember once when a rare Scottish Jubilee, complete with huge St Rollox numbers on her cab worked southwards through Rugby. One hundred and fifty enthusiasts cheered wildly from the tracksides; pens, notebooks and sandwiches flew into the air amid uninhibited enthusiasm.

As young boys, we had our heroes from history and legend, but our true ones were out on the main lines pounding along at a mile a minute with really long snaking expresses—a type seldom seen today. After the demise of steam on fast passenger work much of the railway's romance and atmosphere crumbled and commensurate with the dying steam engine, came a vast reduction in track miles: main lines, cross country lines, branch lines and sidings were abandoned. Engine sheds suffered mass closure—there had been five hundred in 1950—and, long before the steam engine actually disappeared, the tension and magic had been dissipated from train spotting and railway alike.

To new generations, steam locomotives were not greyhounds but sedate plodders. Today, high speed steam has been replaced all over the world—even in developing countries—and although many locomotives remain, they are largely relegated to secondary work. Tragically the days of great train watching are also extinct. Nowhere in the world today can one sit by a busy trackside and see a wide variety of steam trains pass embracing twenty or more different classes. Such days live only in the minds of those privileged to have experienced them; they remain amongst our most treasured memories.

One remarkable exception survived in summer 1977; the high speed expresses running behind the original Reichsbahn 01s between Berlin and Dresden in the German Democratic Republic. With loads approaching five hundred tons, these superb Pacifics were scheduled on mile a minute timings and some truly amazing runs were made by enthusiastic crews conscious of the traditions they were upholding. Imagine such performances in 1977, with Pacifics pre-dating Fowler's parallel boiler Scots on the L.M.S., and the G.W.R. Kings!

Introduced in 1925, the 01s were Germany's standard main line Pacific; building continued up to 1938 and the class totalled some 230 engines. After the partition following World War Two, 01s passed to both sectors and thirty-five D.R. engines were rebuilt from 1961 with large boilers, small windshields, box-pok

PLATE 11. *Opposite:* The splendour of the unmodified Reichsbahn 01 is shown in this scene as one heads the 06.37 express from Berlin to Dresden through the soft countryside near Grossenhain. This fifty-year-old Pacific is in charge of a 450-ton train.

wheels and a semi-streamlined appearance; a few were fitted with Giesl chimneys and some converted to oil firing. Fortunately a number of 01s survived in their original form with large smoke deflectors and upraised chimneys; these were the engines responsible for most expresses between Dresden and Berlin. Handsome as the rebuilds are, the original engines are the most pleasing aesthetically and rank amongst the finest express steam engines in world history.

The frequent service included a number of heavy international expresses such as Bratislava–Berlin with Czech coaches or Sofia–Berlin with Bulgarian, Hungarian and Czech stock. In addition, all trains carried D.R. coaches and a red Mithropa dining car. Thus five different liveries might be seen on one train and this, combined with varying coach heights, provided a truly international flavour. Twelve trains per day were diagrammed for steam haulage, the Pacifics being from either Dresden or Berlin Ost sheds; the former had nine unrebuilt 01s; the latter had both varieties: all were hand-fired.

Enthusiasts from many nations flocked to the linesides to pay their respects and a relaxed attitude towards lineside photography by the G.D.R. authorities enabled visitors to congregate freely by the tracksides and indeed to mix with East German enthusiasts; one notable place being the famous bridge on Weinböhla bank. But unless one's German or French was good, communications were not always possible.

Several trains were scheduled at start to stop averages in excess of 60 m.p.h. over the 103-mile journey between Dresden Neustadt and Berlin Schönefeld, and in summer 1976, the 20.21 ex Dresden was allowed 52 minutes for the 57.9 miles onwards from Doberlug Kirchhain. The 01s were maintained in perfect mechanical order and complete masters of their task but inevitably speculation occurred as to their future. Some reports said steam would survive until electrification—planned for 1980—others maintained the line would be fully dieselized prior to this. At the time of writing, it appears that 1977 will be the last year. By October reports had flooded in that diesels had seriously infiltrated the steam diagrams and it seems too much to hope that the 01s will make a comeback to the June–July level of twelve expresses a day.

The log opposite will clarify the brilliance; it was recorded during August 1976 by Andrew Smith and Roger Blundell. The engine was unrebuilt 01 No. 2207 with ten vehicles weighing 440 tons gross.

A fine example of loco. performance. The net start to stop average was almost 70 m.p.h., based on a net time of $88\frac{3}{4}$ minutes.

This log was jubilantly printed in the September edition of

MILES	STATION	TIME	M.P.H.
0	Dresden Neustadt	00·00	—
1·7	Dresden Pieschen	03.52	45
2·8	Dresden Trachau	04.59	50
4·1	Radebeul Ost	06.26	57
6·3	Radebeul West	08·37	68
8·9	Neucoswig	10.57	64
11·5	Weinböhla	13.32	$57/58\frac{1}{2}/53$
16·8	Bohla	18.54	$71\frac{1}{2}$
21·5	Grossenhain	22.29	$81\frac{1}{2}$
25·2	Zabeltitz	25.12	79
27·7	Frauenhain	27.06	$76\frac{1}{2}$
30·6	Prosen	29.21	79
33·1	Elsterwerda	31.23	33 Slack
37·0	Hohenleipisch	37.17	$46/71\frac{1}{2}$
41·7	Ruckersdorf	41·39	$69\frac{1}{2}/74$
45·4	Doberlug	44.41	70
51·1	Brenitz	49.13	$76\frac{1}{2}$
56·2	Walddrehna	53.15	71/69
58·9	Gehren	55.31	77
62·2	Uckro	57.50	$83\frac{1}{2}/86$
66·8	Drahnsdorf	61.11	79
71·0	Golssen	64.25	77/75
74·5	Klasdorp	67.10	76
77·4	Baruth	69.28	$74\frac{1}{2}/70$
83·2	Neuhof	74.07	81
85·1	Wunsdorf	75.38	71/69 Slight signals
89·0	Zossen	78.56	$71\frac{1}{2}$
90·3	Dabendorf	79.58	69
94·4	Rangsdorf	83.22	$70\frac{1}{2}$
96·6	Dahlewitz	85.20	67
97·3	Blankenfelde	85.59	brakes slack to 47/57 max.
103·3	Berlin Schönefeld	92.58	

World Steam and accompanying it were the comments, 'drivers in East Germany are giving the last real steam expresses a fantastic farewell but they are sometimes exceeding the speed limits. Therefore, please omit dates and train times from your runs to avoid any repercussions.' It was also pointed out that, good as this run was, comparable performances were not uncommon — especially with late running trains and a keen crew.

It is interesting to relate this performance with some British Rail Inter-City services today and taking the loads into consideration, the 01 compares very well indeed. One comparison would be the nine coach expresses running the 99·1 miles from London (St Pancras) to Leicester. In 1977, only one train per day, 'The Master Cutler', was scheduled in 84 minutes, the remainder of the fastest trains taking 88 or 89 minutes. I often travelled on the Cutler and it was usually a very exciting run but, watching the countryside flash by it was heartening to realize that similar performances were still being run with fifty-year-old Pacifics in East Germany.

During our pilgrimage to see them we stayed in Moritzburg—a beautiful castled village—only eight miles from the bridge on Weinböhla bank from which we used to watch the 17.03 off Dresden. It was worked by an unrebuilt 01 often loaded to twelve coaches weighing some 500 tons gross and allowed 103 minutes for the 103·3 mile run. Running like a well-oiled sewing machine, the Pacific confidently swept through the wooded countryside and throbbed up the bank with nothing more than a heat haze rising from the chimney; her huge red driving wheels—6 ft $6\frac{3}{4}$ in. in diameter—spinning gloriously. Flocks of linnets would rise from lineside bushes as the express swept northwards and wafts of oily sulphur drifted across the banks; this combined with the scent of summer evenings, made an elixir of youth; a resurrection of something held dear by innumerable people. Our spirits, though elevated, were tinged with sadness; this might be the last summer: we may never see a Pacific at speed with a heavy express again. We always returned to Moritzburg in time to see a fussy 99.17 class 2-10-2T leave with the narrow gauge passenger to Radeburg. A greater contrast in driving wheels could hardly be imagined; the 01's dashing flightiness and the 2-10-2's shuffling indifference.

We spent an unforgettable day alongside the Berlin–Dresden main line. We left Moritzburg early on a glorious summer's morning and drove through the still countryside towards Grossenhain. Two early morning expresses left Berlin behind steam; one at 06.37 booked non stop in 101 minutes and another at 07.39 which stopped several times and took just under two hours. We reached our location in time for the first train. The banks and field edges were ablaze with colour; cornflowers, wild blackberry and poppies grew profusely highlighting the unspoilt countryside northwards from Dresden: Plate 11, shows 01 2118 heading the non stop past wayside poppy banks. The 07.39 was diagrammed for a rebuilt 01 which used to return with the Pannonia express at 13.32. This was a through train from Sofia which ran up to Berlin in 108 minutes.

Shortly after the rebuilt passed, the 09.29 off Dresden was pegged; this was the Bratislava–Berlin behind an unrebuilt and with a lovely whirling roar from the engine it swept past at very high speed, the coaches screaming across the rails. Only nine steam expresses were booked to pass in daylight but invariably one or two would be diesel. The mid-day Pannonia was usually

PLATE 12. *Opposite:* An East German 750 mm. gauge 99.17 Class 2-10-2T leaves Moritzburg with a passenger train from Radebeul Ost to Radeburg.

late; sometimes by over two hours but its majestic passage was unparalleled and the cavalcade of multi-national coaches swept by as if the steam age were at its zenith.

During the quiet lulls we rediscovered those timeless summers; equipped with cameras, notebooks and apple juice we disappeared into the fields all day. Swallows darted across the meadows and a skylark rippled happily above against a clear blue sky; summer's glory was all around. But later that afternoon there appeared over Grossenhain an inky blue haze backed by ominous storm clouds. Gradually the haze turned white as rain fell in the distance; a meaningful breeze agitated the grasses and the sun went out as if a light bulb had been smashed. The landscape turned into a monotone of subdued hues, its pastoral radiance utterly transformed. The skylark's melody ceased and the bird dropped unobtrusively to earth: no swallows were to be seen. A grey half light heralded the first cold blobs of rain on the heightened breeze and trees—which minutes previously had rippled in golden sunlight—swayed uneasily against an evil looking sky. We darted for cover. As the storm's intensity increased, a whistle sounded and a 52 class 2-10-0 approached with a southbound goods. The Giesl chimneyed giant looked splendid against the storm, her steam and smoke mingling with the swirling elements as if the train were another facet of nature's miraculous drama. We watched her pass through a screen of rain; the long plodding freight standing clear against the skyline. Almost prophetically the skies over Grossenhain began to clear and the passing waggons unfolded a brilliant horizon.

In nearby fields at Bohla were two apple trees and a disused plough which reminded me of the one visible in 'Rain, Steam and Speed'. The afternoon sunlight, now shining clearly, lit southbound trains from the side and enabled the plough to be included in a picture. Only one afternoon express passed behind steam, the 15.22 from Berlin. Just enough exhaust animates the picture; fortuitously, since the 01s run for miles on end with clean chimneys. We had watched many expresses attack the long climb through Weinböhla, but none had emitted visible exhaust—a combination of fine coal and experienced enginemanship.

The rain had added clarity to the atmosphere and, as evening advanced, we heard the rhythms of the 16.48 Berlin–Bratislava as she left Grossenhain some four miles away. A lovely hollow reverberating sound echoed over the damp countryside; for three glorious minutes we listened to the musical roar of an 01 gently picking up speed before she reached our embankment and, with speed in the seventies, she whistled for Bohla station and prepared to make a fast dash down the bank into Dresden.

Nostalgically we made our way back to the road; the semaphore clanked back into place; a perfect day had ended.

Dresden, home of the unrebuilt 01s, is a new city born from the ruins of February 1945. Amongst the city's streamlined modernity lies the awesome remains of bombed buildings; the decaying magnificence of their elaborate facades recalling a past civilization—rudely terminated by the Third Reich. Statues and gargoyles blackened by fire and air pollution, recall many past turmoils and with crumbling expressions gaze open-mouth at the National Socialist wonder around them. These reminders of old Germany, so poignant and stark, stand out like an island of ruined monuments amid a vast ocean.

Bushes and trees have grown throughout the crumbling masonry and querulous jackdaws circle Baroque towers, adding some piquancy of life to this living death. One remembers the songs from the Third Reich; its parades; the specious ideals; the arrogant magnificence of a great nation. Over the road two ballet dancers practised in the open air, as if dancing out a tragedy against a backdrop so searing as to be inimitable on stage. Groups of bystanders watched; the music gently floated across the courtyards and out into the calm city beyond. A steam train whistled in the distance; an 01 headed across the long girder bridge spanning the Elbe and a nineteenth-century paddle steamer hooted as if in salute. As the fifty year old Pacific eased its Berlin Schnellzüge away, it became the living emblem to those acrid ruins nearby.

5

A Glimpse of German Narrow Gauge

When Germany was divided, most of the country's narrow gauge lines passed to the Democratic Republic. All were incorporated into the D.R. but their individualism survived. Since that time, closures have been legion and many fascinating lines have disappeared but East Germany still offers a wide range of systems which provide that unique atmosphere only found on the narrow gauge. And—unlike other countries in the eastern-bloc—the railways can be enjoyed in a relaxed and trouble-free way. Gauges ranging from 600 mm. to one metre exist in settings as varied as the vast plains in the north, the mountainous areas to the west and south, and forest regions of the east.

Motive power extends from small 0-6-0Ts to powerful 2-10-2Ts and classes dating before the partition intermingle with new standard designs built since 1950. All are tank engines of distinctly Germanic appearance; however, three particularly notable types survive: the Saxon Meyer, the Mallet, and the Feldbahn.

The Meyers are four-cylinder compound 0-4-4-0Ts with cylinders placed at the inner ends of two articulated power bogies. The type was widely employed by the Saxon State Railways—both on standard and narrow gauge lines—where engines were needed to negotiate tight curves and steep gradients and, ninety-six 75 cm. gauge Meyers were supplied by Richard Hartmann of Chemnitz between 1892 and 1921. Few remain today, but they do monopolize all traffic on the twenty-three kilometre line running between Wolkenstein and Jöhnstadt on the Czech border.

From Wolkenstein, the line twists its way through a picturesque valley and constantly crosses the river on slender bridges. Fishermen stand patiently in the fast-moving waters; woodlarks sing from the forest edges, and a succession of quiet, sleepy villages line the route. A regular passenger service operates along with a daily freight consisting of main-line

PLATE 13. *Opposite:* The early morning train from Gernrode to Alexisbad headed by a D.R. 0-4-4-0T Mallet, storms past cuttings strewn with the flowers of high summer.

55

waggons running on 75 cm. gauge transporters, and the enormous waggons towering above the diminutive Meyer as it wheezes its way up the valley is one of the most fascinating railway sights in Europe. Very long draw bars are employed to prevent the waggons from touching on curves and the entourage is completed by a small 75 cm. gauge brake van at the rear. The overhanging freight waggons can be a danger—as having become accustomed to narrow gauge clearances—one can easily be taken unawares especially when standing at the trackside preoccupied with photography or tape recording.

Nearby, a Saxon Meyer, along with an old coach and van, has been preserved on a plinth built over the disused trackbed at Geyer. Surrounded by flowers, she provides a fine monument to the rural transportation of a bygone age. Also in Geyer I found a shop displaying model locomotives covering many famous D.R. types—a reflection of the considerable railway enthusiasm in East Germany.

Amongst the scenic beauty of the Harz mountains is the twenty-five kilometre long 'Selketalbahn'—perhaps the most delightful line in the G.D.R. It is graced by the last Mallets working in Germany; six metre gauge 0-4-4-0Ts dating from early this century. This line begins at Gernrode and passes through deeply wooded country to the little junction of Alexisbad. Here the line splits; one section climbing through the hills to Harzgerode; the other meandering through lush meadowland to Strassberg. Trains are frequent and the Mallets work turn about with a unique 2-6-2T; and a 1914 built Henschel 0-6-0T—an oddity with tiny driving wheels and a subsequently top heavy appearance. Whenever narrow bridges or road crossings are encountered, the D.R.'s narrow gauge engines sound their steam bells; these are rung with a convincing resonance by a metal arm which visibly clouts the bell with that comical movement one associated with figures on steam organs.

Alexisbad is a quiet haven overshadowed by wooded hills. The line is entirely steam worked and attracts many visitors—on certain days in the summer, groups of enthusiasts congregate on the platform ends. It all seems too good to be true, yet the line does appear to have a future and during 1977, the old wooden sleepers were being replaced by steel ones.

Another metre gauge system is based nearby at Wernigerode. This is the Harzquerbahn, a heavily graded line sixty kilometres long running through the Harz Mountains to Nordhausen. The line is busy; trains are little short of standard gauge proportions and are worked by sixty-five ton 2-10-2Ts built between 1954/6 at the old Orenstein & Koppel factory at Babelsburg in Berlin; now named Karl Marx Works. Parts of the Harzquerbahn run

PLATE 14. *Opposite:* . . . and another scene at the flower banks on the D.R.'s Selketalbahn, this time featuring the unique metre gauge 2-6-2T No 99-6001.

PLATE 15. *Following pages:* The last German Mallets now perform on the metre gauge Selketalbahn in the Harz Mountains of the G.D.R. Here we see one of these turn of the century 0-4-4-0Ts leaving Gernrode with an early morning train to Alexisbad.

close to the West German border, particularly the branch to Schierke; another branch runs up to Hasselfelde. Together, these two systems—which are not connected but have a common works at Wernigerode—provide sixty route miles of metre gauge with engines ranging from seventy-five year old Mallets to modern 2-10-2Ts working amid the splendour of the Harz.

The last German 'Feldbahn' 0-8-0Ts can be found near the Polish border working the 'Muskauer Waldeisenbahn'—a 60 cm. forestry railway based on Bad Muskau. These were once standard engines for military field railways and colonial lines. Over two thousand five hundred were built by eleven different makers and though detailed variations occurred within their ranks, all were very similar. The Feldbahn proved a strong engine for its size; it had excellent adhesion and the eight-coupled wheels were made flexible by using Klein-Linder axles front and rear. Until recently, a stud of Feldbahns worked the maze of 60 cm. lines which proliferated around Bad Muskau, but sadly most have now closed down and at the time of writing only two engines were being steamed daily, the remainder standing derelict.

This line runs along the banks of the River Neisse and on the opposite side lies Poland. After many adventures searching the woods—only to discover derelict sidings—we found a live Feldbahn working a clay pit near Halbendorf. According to her crew, the engine's boilers were rapidly wearing out and it would only be a matter of time before these celebrated little engines became a memory.

Another lovely 75 cm. gauge line runs from Radebeul Ost in the suburbs of Dresden out to Radeburg, seventeen kilometres away: it has a lively passenger service plus one freight a day and is operated by modern 2-10-2Ts from Karl Marx Works. The route crosses one lake on a causeway, skirts another, and passes the castle of Moritzburg. On certain days, Saxon Meyers work enthusiasts' specials. This system is easily combined with visits to the Berlin–Dresden main line, Radebeul being a high speed point for Dresden-bound expresses and even when the Pacifics have gone, there will still be the paddle steamers on the Elbe.

The demise of German narrow gauge continues, but at present, sufficient interest remains to warrant a visit, especially when this is combined with the excitement of the standard gauge, where two- and three-cylinder Pacifics, 2-10-0s, and 2-8-2s all remain vigorously active. The country provides a haven of delight for European enthusiasts and should continue to do so for a few precious years to come.

6 The Little Fireless of Schwertberg

This remarkable engine had intrigued me for years and I determined to fulfil an ambition to see her. Records showed her as an 0-6-0 Fireless employed by the Schwertberg Works of Kamig A.G. in northern Austria. Her duty was to carry trainloads of prepared clay along a 600 mm. gauge branch to a connection with the O.B.B.'s St Valentin–Krems line. Diagrams showed the engine to be fascinating, with two outside cylinders placed at the front, and a conventional engine's chimney—aspects almost unknown amongst Fireless locomotives. Unique in appearance, she looked like an enormous Meccano-plated toy and yet her shape was strangely ethereal—rather like the 'Firefly' in Turner's 'Rain, Steam and Speed'. In brief, I had to go and see the engine for myself.

With great excitement, we set out from Linz and, travelling through rich agricultural countryside, headed for the small village of Schwertberg. This part of Austria was not new to me, several previous visits having been made to the Linz–Summerau line—one of the last to retain standard gauge steam in Austria. In those days, Fireless engines had tended to be overlooked and although the passing of main line steam was a tragedy, it has enabled many of us to appreciate steam railways from a much wider spectrum, especially the fascinations offered by industrial lines.

Schwertberg proved unexpectedly picturesque. The River Aist flows sedately through the village centre and is crossed at intervals by ornate bridges. Flowers bloomed in profusion, and a fifteenth century castle benignly overlooked the narrow streets. This tranquil place seemed an unusual setting for an industrial railway; but there, running down the main street, were the shiny metals; passing within a yard of superbly maintained residences with their elegant shutters, colourful window-boxes and brass door trimmings. It was lunch-time; hardly a soul was about.

In an effort to find the engine, we followed the track to the

O.B.B. connection. Upon arrival, we found several lines of China-clay waggons but no engine. 'The "Wursteldampfer" has gone back to the plant', a workman said. Setting off uptrack, we pondered upon his description; we were to learn that 'Wursteldampfer' means 'Sausage Heater' or 'Hot Dog Machine', the name by which the little Fireless is locally known.

Having left the village and its castle behind, we followed the roadside line as it curved through delightful wooded country. The track bed was strewn with acorns; apples, pears and damsons grew abundantly in adjacent orchards; and the golden sunshine belied summer: autumn was on its way. On our right, the Aist bubbled over a stony bed, with Mallard taking refuge in deeper pools beneath the banks. At intervals there grew tall yellow Aist flowers; a wild beauty only found in Schwertberg and one other place in Asia. Jays screeched in the woodlands around—as continuing our journey—we came upon a cat's grave in the embankment. It was a beautifully tended mound of soil, free of weeds and graced with two crimson roses. The face of a black cat was painted on an improvised headstone made of tree bark; some child from the village must have found solace in this place. Indeed, Schwertberg seemed more like a fairy grotto than an industrial village, but further up the valley we caught a glimpse of the four kilometre long cableway which carries unprepared clay from the field to the plant.

The roadside works merged with the hillside and a rake of waggons was being filled from overhead hoppers. Immediately behind the track ended. Where was the Fireless?—doubts began to well up. Suddenly, she burst from a dark shed and, with steam oozing from both cylinders, shuffled down to the waggons and prepared to couple up. She had just taken a change of steam from the work's boilers, and being fully rejuvenated, was raring to go. There was little time to bestow admiring glances; a whistle cry broke out, and without further ado, she picked up eight loaded waggons and headed away down the valley. Pure white steam billowed after her, like snowballs flung from her chimney.

She was built by Floridsdorf of Vienna in 1930 as their works No. 3012 and came to Schwertberg a brand new engine. The plant commenced operation in 1922 and initially employed horses and waggons to carry the prepared clay; when the rails were laid, they followed the same route in the manner of a roadside tramway. Towards the end of World War Two, horses were again used for a time and the Fireless hidden away in case she might be seen and destroyed by the occupiers. The engine is in my experience unique although it has been suggested that a sister was built by Floridsdorf during World War Two, apparently for use abroad.

PLATE 16. *Opposite:* Kamig AG's delightful 0-6-0 Fireless takes a breather from hauling China Clay and stands in repose at the Schwertberg exchange sidings with the Austrian Federal Railway.

Schwertberg's residents accept the engine as if it were nothing unusual; certainly it has been in the village as long as most can remember. Several people I met were surprised, if not actually disbelieving, when I mentioned their engine had no fire inside. 'But it's a steam engine', one man said loftily—evidently regarding me as too young to remember them. No one seemed concerned when the little train criss-crosses the road and chugs past the front of houses, a modern pharmacy and supermarket—incongruous as this might be—but chagrin is expressed about the discoloration of the Aist by the clay washing plant. This village of four thousand inhabitants has a further claim to fame; the home of Austria's oldest inhabitant—a woman of 105 years old.

The driver suggested that we join him at 5.00 the following morning to see the engine take her first charge of steam. The factory boilers operate at 16 Kg cm.^{2} and the engine is fed through a 'plug in' connecting valve up to a pressure of 10 Kg. cm.^{2}. When working, steam is fed into the cylinders at a lower pressure through a reducing valve; this ensures a constant power output, and until recharging is almost due, avoids any fall off in performance. When starting from cold, $1\frac{1}{2}$ hours are needed to fill up, but having become 'warm' successive charges take only half an hour. With a full boiler, the Fireless becomes $3\frac{1}{2}$ tons heavier. One charge gives the engine 114 h.p. and enables her to take a fourteen waggon train, weighing sixty tons, down to the O.B.B. four kilometres away—shunt the exchange siding—and return with a rake of empty waggons: a round trip of forty minutes. Care must obviously be taken lest she runs out of steam somewhere along the line—a particular hazard in cold weather when the pressure falls more rapidly.

The present driver has worked on the engine for twenty-two years, during which time several unfortunate incidents have occurred. He told us that derailments and icy rails could invariably be relied upon to deposit the unhappy engine in an embarrassing situation without steam; the problems of returning her upgrade to the factory being legion. Few engines have the animated personality of this Fireless; she is the epitome of a Rev. Audrey creation. One can imagine her as 'Franz the Fireless' having come to a standstill on the road crossing one blustery snowy day. No steam remains and Franz, to the despair of his driver, prepares to sleep; the traffic queue becoming longer and longer!

Although her misdemeanours have been minimal, she asserted her independence one night by running away. It was during a late shift; the engine had been filled in readiness for a trip, but when the crew came to take her out she had gone. Pandemonium broke loose, the sidings were searched to no effect: the engine was not at the factory. A search was made of the line, but she had vanished into thin air. A panic stricken phone call to the village elucidated the information that she had passed through half an hour since—narrowly missing a car—but no one had realized the engine was not manned. The crew, along with a fitter, set off for the exchange sidings, horrified by what they might discover. For five minutes, lamps flashed around the darkened sidings until, there in a far-away corner—up against a mound of earth—she stood; like the Titfield Thunderbolt itself. Although none the worse for wear, the steam pressure had fallen too low, preventing her from returning to the factory. Small wonder everyone gave it up that night and repaired to the local ale house.

At present, workings are confined from 6.00 a.m. until 15.00 on week-days only. Having reached the plant by ropeway, the unprepared clay is washed to remove any sand particles and later pressed to remove water. It is then dried on a conveyor belt by high pressure steam before despatch in either powder or nugget form. Some 800 tons of prepared clay are carried weekly and each day the Fireless makes about five journeys to the O.B.B. Schwertberg's china clay goes to Italy for use in paper and paint manufacture, although a little finds its way into medicinal tablets.

Today, Fireless engines are almost exclusively yard shunters; it is very unusual to find one working a branch line. Paradoxically, some of the first examples built were for passenger carrying tramways and, as long ago as 1873, a small Fireless ran successfully down a $3\frac{1}{2}$ mile branch line and back in the suburbs of New York. The engine started with 180 lbs p.s.i. and returned with 45 lbs p.s.i. Others were built in America and Europe.

For industrial use, the Fireless offers two distinct advantages: firstly, absolute safety for users with a fire risk—such as munitions works and paper factories—where sparks emitted from a conventional engine could wreak havoc; and secondly, it constitutes an extremely low cost shunting engine for users with a ready supply of high pressure steam—such as gas works, power stations and various manufacturing industries. Britain's first industrial Fireless worked for the Thames Paper Mills and though imported from Germany, it was a judicious purchase as the company had a fire risk and abundant steam!

Prior to World War One, Fireless engines were active in Europe, especially Germany, where they were pioneered by Orenstein and Koppel. Britain's first industrial example did not appear until 1913 when Andrew Barclay of Kilmarnock built one for an explosives factory. The ensuing war, and subsequent upsurge in the production of munitions created a demand and more were built; many passing to other industries when the

PLATE 17. *Following pages:* One of the most delightful engines extant in Europe today is this 0-6-0 Fireless built by Floridsdorf of Vienna in 1930. She works for the Kamig AG China Clay plant at Schwertberg in Austria.

hostilities ended. Andrew Barclay were to become the leading producers of the Fireless engines in Britain and between 1913/61 they built one hundred and fourteen examples. This figure exceeds the Fireless output of all other British builders combined; the national total being one hundred and sixty-two.

Some survive in Britain, but West Germany, having made better use of the type, has some really potent examples in operation. Indeed, Fireless engines are still being developed there; some having been built during the 1970s. In lamenting the cessation of steam building throughout the world, we often forget the Fireless—in some ways the most efficient shunting engine ever devised. It is amazing that industrial plants with steam available should go to the expense of buying and running diesel locomotives, when for a much lower capital outlay, a Fireless engine—guaranteed of longevity—could be used at minimal cost.

But the future of Schwertberg's engine is sadly limited. Kamig A.G. have formulated a plan to dispense with the aerial ropeway and railway. They intend to wash the clay in the hills and pipe it directly to a new drying and processing works next to the O.B.B. at Aisthoven—one station east of Schwertberg. Construction of the pipeway has already begun.

There are hopes that Schwertberg's engine can be preserved. Her crew and many villagers would like her to be mounted on a plinth in Schwertberg as an industrial monument. Already a ceremony is being arranged for the last journey; a brass band will be turned out and everyone will pay tribute to the little train that has enriched Schwertberg for half a century. But it will be a sad day when the death knell finally tolls.

7 A Graveyard in Greece

The steam locomotive's personality is both diverse and illusive; many of its charms are obvious enough, but to attempt a neat paraphrase of its allure would test even the most eloquent. But one delightful aspect of the subject is the vast historical canvas from which today's steam survivors are drawn. Like great architectural works, the historical locomotive provides a living link with the past and, in a most vivid way, helps us to see the present, not as a tense and intransigent mood, but as a part of an ever developing history moving towards its own conclusions in the fullness of time. In brief, the steam engine has become a monument and, like the great architectural designs and periods, it too echoes its particular day and age.

I believe this evocation of a continuing history to be one of the most valuable experiences offered by our working preserved lines. They are akin to stately homes as part of a national heritage : the living inverse of the time machine in fiction.

These were my thoughts as I wandered waist deep in scrub through the locomotive graveyard at Thessalonika in northern Greece; it was the end of that country's steam age. Greece had no discernible school of design; her engines were all imported. The resulting fleet was splendidly cosmopolitan, every major locomotive building country being represented. But around these graveyards two important periods of design were evident; firstly some of the last remnants of the great Austrian Empire school and secondly some famous and distinctive types specially produced during World War Two. An unlikely combination indeed.

Perhaps surprisingly, the engines of the former group excited me the most; certainly those from World War Two were more familiar and indeed all had worked in Britain, but I find something awe-inspiring about the great Austrian designs, especially those by Karl Golsdorf—Chief Mechanical Engineer to the Austrian State Railway from 1891 to 1916.

Before the Habsburg Empire was split up in 1919, the Austro-Hungarian section covered large areas of Eastern Europe, and under Golsdorf there developed a remarkable group of steam classes which formed a significant part of Austrian design. During the twenty-five years that he held office, Golsdorf produced over fifty different classes—a figure matched by few engineers in history. But more significant, was the overall soundness in design; combined with the huge building totals to which some of his creations eventually aspired. Several Golsdorf freight classes survived into the 1970s and one remembers how he graduated from 0-6-0s in 1891 to 2-6-0s a few years later; this developed into a 2-8-0 in 1897 and by 1900 a superb 0-10-0 had appeared; and before his tenure of office ended, 2-10-0s and even 2-12-0s had been embraced. His 2-8-0 of 1897 was the first important example of its type in Europe and has been regarded as the forerunner of the present day general freight engine. Classified 170, these two-cylinder compounds eventually totalled around nine hundred examples and a further five hundred were added in simple form after 1917. Known as 270s, these were discussed in some detail in my *Masterpieces in Steam*.

However, the Golsdorf classic which loomed dramatically out of the scrub in that Thessalonika graveyard was the simple version of his 0-10-0 of 1900 and, despite being rusted, they seemed to exude a latent energy. Initially these engines began as two-cylinder compounds with a low pressure cylinder of no less than $33\frac{1}{2}$ in. diameter. Capable of pulling seven hundred tons over 1 in 100 grades at thirteen m.p.h., the breed was originally intended for coal-carrying lines in Bohemia, but soon became popular on a wide basis. Over four hundred were built in compound form, both saturated and superheated—classified 180 and 80 respectively—until, with the general abandonment of compounding, a further four hundred were built from 1916 onwards as simples with piston valves and superheaters: these were known as the 80.9s. Greece ordered fifty from S.T.E.G. in Vienna during the mid-1920s and classified them 'Kb'. They provided the country with a fine standard class of powerful engines with a wide route availability since one of Golsdorf's hallmarks was the design of large conventional engines with a light axle-loading and the ability to negotiate tight curves. Some twenty of these Trojans lay moribund in that sad place and I yearned to see one at work leading a heavy freight. Despite the years of indolence, many looked capable of being up and away for want of a mere charge of steam.

Imagine wandering amongst these abandoned giants with banks of purple flowers up to your waist. In fact, the combination of dense undergrowth and small trees virtually obscured the view

PLATE 18. *Opposite*: Rotting away amid the locomotive graveyards at Thessalonika lies this handsome giant; a 2-10-0 from Skoda built to the design of the old Austrian Sudbahn 580 class. Classified 'La' by the S.E.K.

ahead and although difficult to negotiate they did create a heightened atmosphere of discovery. The awesome presence of decaying engines will be well imagined by anyone who has known such places and throughout my stay I felt a ghostly sensation that engines were watching me. I was reminded of that dreadful night in Salisbury on 9 July 1967 when, after the cessation of Southern Region steam working, all surviving engines were despatched to either Salisbury or Weymouth sheds. By nightfall, Salisbury depot was full of condemned engines all in steam but with their fires dropped; it was like a graveyard come alive, with tormented spirits rising up on all sides amid a macabre cacophony of hissing and gurgling silhouettes. I have had many strange experiences with steam locomotives but that night remains one of the most poignant.

At Thessalonika the warm sunshine illuminated a hundred different tones of rust: ladybirds graced the wild flowers whilst the soft summery drone of bees sedately engaged in collecting pollen, provided a suitable accompaniment to the quiet timelessness of the dump; Plate 18 catches the intrinsic mood.

With all the delight of lineside discovery in childhood, I came upon another 0-10-0 quite different from the Golsdorf strain. It was a member of the Greek 'Kg' class, one of twenty engines delivered from Belgium in 1929. No less than four builders had been engaged in producing these: Tubize, Haine St Pierre, St Leonard and La Meuse. I imagine they must have been almost the last 0-10-0s ever built and were of an infinitely more modern aspect than one normally associates with this wheel arrangement.

But, the pièce de resistance was the Greek 'La', a magnificent 2-10-0 of the old Sudbahn Class 580 built in the finest Austrian tradition. Eight were present in various stages of decay from a class of forty built specially by S.T.E.G. of Vienna and the Skoda Works at Pilzen in Czechoslovakia in 1926/7 respectively. How these Austrian 2-10-0s set one's imagination racing; what magnificent beasts they were. But I had never seen one turn a wheel, many long years having passed since any were active in Greece and none survive elsewhere. And yet I rate them as one of my favourite locomotive classes. The closest I ever came to seeing one in action was during my time in Yugoslavia with Tadej Bratè, whose work *The Steam Locomotives of Yugoslavia*, has become well known in railway literature. We were driving back to Ljubljana one evening after a day's photography, when we passed a goods yard and I caught a glimpse of a gleaming black 2-10-0 with elongated proportions and exceedingly handsome appearance: one of the most splendid apparitions I had ever seen. 'That's one of our last 29 class—Austrian Class 81 of 1920', Tadej said. The yard was out of bounds to us, but Tadej assured me that

PLATE 19. *Opposite:* Another famous type associated with war was the U.S. Army Transportation Corps 0-6-0T. Here one takes refreshment whilst shunting at Drama in Northern Greece. In the post-war period, engines of this type shunted Southampton Docks.

we would find one working another day; he knew of a place. Unfortunately, Tadej was summoned away to work prematurely and I never did see another 29 and the tantalizing memory of that spotless thoroughbred—the ultimate in 2-10-0s—will live with me forever. The Sudbahn 580s in Greece were not the same as Yugoslavia's Class 29, but they were similar, and both had enough of the classic Austrian lineage to set my pulses truly racing. The fall of the Austro-Hungarian monarchy spread Golsdorf's engines over the railways of the various states that emerged; their building continued long after the empire's dissolution, but very few examples exist today either working or otherwise.

It's a shame that many railway enthusiasts orientate so markedly towards the products of their homelands and show little more than a passing interest in other countries' designs. This is surely the ultimate sin for British, German and American enthusiasts since their indigenous railways were but a fraction of the whole: what was good for them often being good for the rest of the world too.

However, had I been in a parochial mood, Thessalonika would not have disappointed me, as a British W.D. 2-10-0 stood rusting alongside a Sudbahn 580. Some thirty years separated the two classes, the Austrian engine being one of Europe's first 2-10-0s and their design, from an aesthetical viewpoint, has—in my opinion—never been surpassed.

Another well-known type associated with Britain is the H.S.R. 'Da' class. Known as the United States Army Transportation Corps 0-6-0T, they are the American equivalent of Britain's Hunslet Austerity and were produced by H. K. Porter, Vulcan Loco Works Wilkes Barre and Davenports for shunting in depots, docks and yards—first appearing in 1942. They became far more widespread than their Hunslet relations, working throughout Western Europe, Greece, the Middle East and North Africa. Over five hundred were built and after the war they remained active in many countries; Yugoslavia obtained one hundred and six through U.N.R.R.A., classifying them 62, and even built a further twenty-three examples between 1956/9 at their Slavonski Brod Works. By the late 1960s, these 0-6-0Ts were still found in Britain, France, Yugoslavia, Greece and Iraq. They were destined to become the final steam class in regular service on Greek railways and some underwent an interesting hybridization by removal of their side tanks and the addition of old tenders from scrapped locomotives. This modification utterly transformed their typical American aura and as I progressed around the dump I discovered them in several different guises.

British enthusiasts will remember them as Southern Region numbers 30061–30074. British Railways acquired this batch from the War Department for shunting at Southampton Docks, working from 71's depot, they lasted until the early 1960s. Today, one can still be enjoyed in operation on the Keighley and Worth Valley Railways in Yorkshire, she is their number 72, once B.R. No 30072 built by Vulcan Ironworks in 1943. No 72 is in good company at Keighley with other war-time designs but has demonstrated her versatility by making a spectacular appearance in the film 'The Railway Children'.

The hours pass swiftly amid locomotive dumps so much is there to see, study and ponder upon, thus afternoon shadows were beginning to lengthen before I discovered two heavily rusted S160s—the great American 2-8-0 of 1942. Although not illustrated in this volume, the S160's working life is inextricably bound with other war-time designs; accordingly they warrant some mention in these pages. Built to the British loading gauge, the class was designed by Major J. W. Marsh of the United States Army Corps of Engineers. Small by American standards, the S160 was specially prepared to meet the very worst kind of route restrictions likely to be encountered in either Britain or Europe and their length was within 61 feet to conform with smaller turntables. Over two thousand were built by the 'big three' American works: Baldwin, Lima and Alco. The first arrived in England in 1942 beginning work on the Great Western and many readers will be familiar with W. M. Earley's pictures of them on freight haulage around Reading. After war-time service, they became surplus to requirements and the Communists bought them in their hundreds; Poland, Czechoslovakia, Hungary and Yugoslavia; Turkey also took a batch. Greece obtained twenty-seven after the war but received a further twenty-five from Italy in 1959 making a total of fifty-two.

Some Greek S160s survived on standby duty in 1976, but most were consigned to dumps; today the type only remains active in Poland and Turkey. During 1977 it was announced that a Polish example had been purchased for the Keighley and Worth Valley Railway: this enlightened preservation further enhances that railway's marvellous stud of war engines.

The S160 is part of a family within itself since two hundred larger examples were delivered to Russia as 5 ft 0 in. gauge engines in 1943 under the Lease-Lend agreement. These formed the 'Wa' class and some survived until quite recently. Others went to India's 5 ft 6 in. gauge lines, becoming classified A.W.C., and it was my pleasure to see several hard at work in 1977 on industrial pick-up freights in the Calcutta area.

I have traced my progress around the dump in words, but the last forlorn hulk encountered was in absolute contrast with the distinctive families already described. She was an enormous 2-10-

2, one of twenty giants delivered to Greece during 1953/4 from the Italian works of Breda and Ansaldo. Built as part of Italian war reparations to Greece, such a project was not Italy's forte; she had barely built a steam engine for thirty years and had no export tradition; furthermore Italy had never built an engine of such proportions before. This indiscretion cost the Greeks dearly, as considerable troubles were experienced and the engines were far from successful; few Greek enginemen having much good to say for the 'grosse machine', stupendous as they were to the eye.

With this evolutionary freak, devoid of family lineage and built out of her time, my spell with history ended. Apart from the thrill of their brooding presence, the engines had taken me on imagination's wing over the old Hapsburg Empire and across Europe in those bitter years of World War Two; times when railways were railways as they can never be again.

A sunset was welling up across the Mediterranean sky as I made my way from that sad place back to the liveliness of Thessalonika; but I could not resist occasional backward glances at the stark array of silhouettes against the sky for they were outlines that, over the years I had come to know and love. Their memory lives within me.

8

By the Lineside at Izmir

With several scruffy urchins in tow, we fought our way through Izmir's thronged bazaars. On either side, tightly packed stalls glistened with every conceivable kind of merchandise and a hundred calls in brazen Turkish begged us to buy. Had our mission been less urgent, we may well have done so, but with all the ardour of crusaders seeking the holy grail, we continued in the direction of Hilal Railway Crossing—one of the world's last great train watching places.

Izmir, set on Turkey's Aegean Coast, is the country's second seaport. Traditionally famous for Smyrna figs, this bustling city now has steam locomotives added to its fame; the busy suburban services which operate from the two terminal stations, Basmane and Alsancak, were until 1977, almost exclusively handled by vintage steam engines. Less than a mile out, the two lines cross one another on the level at Hilal where they are overlooked by a tall and imposing signal box known as 'Poste B'. A lovely sooty smell enveloped the crossing and I marvelled at the gleaming silver rails polished only by steam engines; here was a revered place indeed. Sheep grazed on every grassy tuft, cared for by ragged boys a mere eight or nine years old, who, finding us some relief in their hours of indolence, stared unashamedly in our direction. The signalman gave a cheery wave; he had seen the likes of us before; indeed, all along the track sides were pieces of film cartons left by ever increasing bands of international enthusiasts.

A signal clanked upwards and heavy smudges of black smoke could be seen rising from the direction of Alsancak station; a passenger train was on its way. Mesmerized, we watched the train approach, what class was it? We would have identified any class immediately, had we been in Britain twenty years ago, but our eye practice for foreign engines was not developed, familiar as many were from the pages of books. 'She's Prussian', I said—not a profound observation—I confess—but it's surprising how

PLATE 20. *Opposite:* One of the T.C.D.D's ex Prussian G8 0-8-0s makes heavy weather of the climb from Izmir with an afternoon train to Buca.

tantalizing locomotive identification can be. With her acrid smoke palls undiminished in their glory she bore down upon us and her whistle opened up emitting a thin screaming wail for the crossing approaches. No. 44071 I read from the smoke box as she bustled past; a Prussian G8; a cornerstone in steam design and a member of the third most numerous class in world history.

After this excitement we omitted to see a peg lift on the Basmane line and were surprised by a six-coach train from Ciyli headed by a gleaming ex-Ottoman Railway 2-8-2 No. 46102 running tender first. Pride of her crew, she was bulled up in black livery with red wheels and gleaming brasswork in the best of pre-grouping traditions. She was British to the very last detail, although her wheel arrangement was 'strictly for export'. Six of these delightful engines had been sent to Turkey from Robert Stephenson's Newcastle Works during 1929/32. Her number merits further mention; many will recognize it as identical with Royal Scot class 'Black Watch'—that rare beast from Polmadie Glasgow.

To witness an Ottoman 2-8-2 at the bufferstops amid Alsancak's gloomy portals was a fine experience. It might have been a British terminal station in the early 1930s, for the engines—as English as cheddar cheese—and almost as graceful as a Great Western Castle, produced a fine period atmosphere and one almost unwittingly looked around for advertisements offering Virol, Craven A or Reckitts Blue.

Alsancak was the headquarters of the British owned Ottoman Railway Co.; their line from Smyrna (Izmir's original name) to Buca was the first in Asiatic Turkey and was opened in 1860. Extensions eastwards to Egridir facilitated heavy traffic in cotton fruit and corn from the fertile valleys. Alsancak provided access to Izmir Docks which, until the mid-1960s, were shunted by some curious flat-topped 0-6-0STs first introduced by Stephensons in 1875. The O.R.C. remained in British hands until 1935, when it was incorporated into the T.C.D.D. Today, suburban services extend to Buca and Seydikoy with long-distance trains to 'O'demis, Denizli and Afyon.

Basmane, on the other hand, was headquarters of the French-owned Smyrna Cassaba et Prolonguements Railway. Originally British, this had been sold to the French in 1893. It extended northwards to Bandirma on the sea of Marmara and eastwards to Afyon. This network passed to the T.C.D.D. in 1934 and today the suburban trains from Basmane embrace Ciyli and Bornova, with long-distance workings to Manisa, Bandirma, Eskisehir and Ankara the Turkish capital.

Izmir's vintage motive power can be traced back to these companies, although nowadays no attempt is made to restrict the

PLATE 21. *Following pages:* One of the T.C.D.D's magnificent German built 2-10-2s of 1930s vintage, storms the heavy grades amid the teeming suburbs of Yesildere with a mixed train from Izmir to Denizli.

engines to their former lines. We were delightfully reminded of this fact when, less than twenty minutes after 46102 had passed, a 2-10-0 from Corpet Louvet—with the highly un-British number of 56912—went charging over the crossing with an Alsancak to Seydikoy train. The S.C.P. took eight of these lightly loaded Decapods from the Paris Works in 1926/7 and for so large an engine, they possess a stimulating air of vintage.

Having witnessed some of these remarkable performances, we determined to return at sunrise the following morning for a day-long session. It was to be one of the most memorable of my life.

Shortly after passing Hilal, the Alsancak line begins to climb sharply through the teeming suburbs of Yesildere and curves upwards against a stunning backdrop of houses, mosques and minarets perched upon every conceivable inch of a vast hillside; this was our chosen location for that unforgettable day. Although we would miss the Basmane trains, motive power on the two routes was mixed, if not actually interposed, as all engines came from a common shed at Halkapinar. The backdrop and curved climb provided fine opportunities for picture making from vantage points on the grassy slopes and rocky defiles immediately alongside the track.

Our minibus to Yesildere had a seating capacity of twelve, but this did not deter our driver from packing in twenty-three no less; indeed he showed little interest in moving off at all until his charges had exceeded twenty in number. Although no one spoke a word of English, our fare was paid for us and we emerged the richer by three sticks of chewing gum, two oranges and a bunch of grapes.

Leaving the road, we climbed towards the line and breathlessly reached track level just as the sun rose over the hillside; within minutes the track bed and Yesildere were engulfed in a radiant glow. Almost immediately, a steady panting rhythm heralded the approach of Humboldt 2-8-0 No. 45132 with the 7.30 train to Buca. French engines are rare today and these survivors from the S.C.P. are especially interesting. Delightfully French, a dozen were built by Humboldt of Paris in 1912 to a design originally prepared by Maffei in 1910 for the Damas Hama et Prolonguement in Syria. A little earlier, the 6.40 to 'O'demis had passed behind a Stephenson 2-8-2, the only regular working for this class from Alsancak.

Having made our lineside base, we studied the timetable and wondered what engines would be working the trains scheduled to pass. Next was the eagerly awaited 7.50 long-distance train to Afyon which ran thrice weekly—usually behind a D.R. class 52 2-10-0 'Kriegslokomotive'. A tenseness always precedes the arrival of an important train and this became evident as we

PLATE 22. *Opposite:* Another French locomotive! An impressive Corpet Louvet 2-10-0 of 1926, caught storming through Yesildere with a suburban train to Buca.

PLATE 23. *Opposite*: Another classic design in world history is the ex-Prussian G10 0-10-0 T.C.D.D. 55001 class. This twilight study shows one preparing to leave Afyon with a goods train.

listened, straining our ears for the slogging exhaust beats as she attacked the bank. But now Yesildere's populace were up and about, and the general clamour from the hillside dwellings had to be heard to be believed. Eventually the war engine became audible, as running fifteen minutes late, she bit into the grade in fine style. Seconds later, a Prussian G8 running tender first went down the bank with a local from Seydikoy. These two classes are the best known German designs and are amongst the most numerous steam types built in world history. The Kriegslokomotive, a barren Austerity built to follow Hitler's armies during World War Two, eventually totalled 6,300 engines; whilst Prussia's G8 exceeded 5,200. One marvels that, sixty-five years after their inception in 1913, these Prussian goods engines should be working passenger trains around a modern Turkish city; it was thrilling to see both classes pass simultaneously.

Next in succession was an 09.00 departure from Alsancak to Buca, a train invariably worked by diesel railcar—thus attentions were now focussed on the 09.35 mixed for Denizli. This famous train often produced spectacular double headed combinations, such as Corpet Louvet 2-10-0s and German 2-10-2s; or either of these, with a Prussian G8 or G10, possibilities being limited only by the classes found at Izmir.

By this time, we had been at Yesildere long enough to have attracted attention from local children and we became surrounded by a chattering group. Alman, Ingliz or American?— demanded their leader with his ten words of broken English. Our presence mystified them, for Turkish children have little regard for steam engines, although when something abnormal happens—as it did later that day—they will respond with uninhibited delight. A group of gum-chewing soldiers sweltering in thick uniforms and complete with rifles and crew cuts came over to stare at the foreigners and their sophisticated cameras; the blank incomprehension of their gazes clearly revealed an ignorance of both photography and railway enthusiasm; unnerving as these riveted attentions were, we were meant no harm.

Old Hannah soared higher; the day's blazing heat had begun. It was now 10.30. The Denizli was late. Conjecture on what kind of double header we might see was shattered when the mosques began to wail in garbled Arabic; first one, followed within seconds by another, until five were in full cry, rising and falling in a tuneless cacophony of sound to call the faithful to prayer. Although falling somewhat hard upon western ears, there is beauty in these eerie emanations which mark important divisions in the Moslem day. But soon the rather more harmonious sound of a climbing train was heard, as palls of exhaust rose over the hillside. A timely drift of cloud dominated the skyline yet the sun beamed with all its clarity from a clear sky; everything was set for a fine picture. The Denizli was lightly loaded, only one engine being provided—a splendid German 2-10-2. Twenty-seven were built for the T.C.D.D. between 1933/7 by Henschell, Krupp and Schwartzkopf; a powerful lightweight design in which the German builders skilfully utilized the boiler of a T.C.D.D. 4-8-0 class with the wheels and motion of a Prussian G10. Turkish rail travel is slow; the train being allowed $11\frac{1}{2}$ hours for the hundred and sixty mile run to Denizli.

This excitement heralded a quiet spell until lunch-time and the soft grassy banks provided a glorious repose. Indeed I must have dozed awhile, for I awoke with a start; a steam engine was quietly panting its way towards us. Running light and puffing sedately, a gleaming black engine rounded the bend: Humboldt 2-8-0 No. 45131. She had been outshopped from Halkapinar Works that very day and was 'running in'—a 1912 built 2-8-0 in 1976, marvellous! An aroma of fresh paint wafted across the banks as she passed: immaculate in every detail, she had received a major overhaul. I noticed that she differed from No. 45127 in having a rather more stubby chimney: doubtless the original one was lost years ago.

Our next train was the mid-day to Buca—presumably Humboldt No. 45127 returning—though such logical continuity did not always apply. It was in fact a grimy G8, which in spite of an ailing appearance, lifted the five-coach train over the drag with a completely clean chimney; nothing more than a heat haze being visible. There were some amazing contrasts in the trains; one would struggle up like a volcano on the move, the next would pass in the manner of this G8. Within minutes a Seydikoy train followed headed by one of those handsome Corpet Louvet 2-10-0s which romped past in fine style, its small driving wheels biting confidently into the adverse grade.

Nothing further was due for two hours, and so we momentarily left our vantage point, crossed a rough meadow strewn with sheep's skulls and reached the outer buildings of Yesildere. Figs and golden brown pomegranates grew abundantly in wayside groves but our mission was to indulge in one of Turkey's most underestimated delicacies—peaches, the most perfect specimens imaginable being freely available. Loaded with our bounty, we returned to the trackside and sat under a spreading tree; it seemed as if the whole of Yesildere was taking a siesta; insects droned sleepily and a molten heat haze oozed upwards from the tracks. The world was at peace.

Gradually, I became aware of a series of muffled bangs. The

beats were uneven and though distant, echoed throughout the valley; sometimes they died away completely, only to recur after intervals of several minutes. It was an uncanny sound which seemed to reverberate from the earth itself, like some terrestrial explosion. I thought of Turkish carnivals and all their strange appendages and imagined a giant in gawdy costume striking a tightly stretched snare drum of enormous proportions, followed by a hundred ghoulish Turkish children apeing his every action. Having seemingly died away, the noise would suddenly return closer than before. By now, I was on my feet scanning the hillside for any clue, but all went quiet; only the gentle cackle of humanity amongst the myriad dwellings being audible. No sooner had I sat down than an explosion rang out from towards Izmir, followed by a rapid succession of high-pitch bangs which had a strange familiarity about them: it sounded like an engine in great trouble. The roars, bangs and periods of silence continued until, far behind Yesildere, smudges of smoke became visible.

But it cannot possibly be a train struggling up the bank, I thought; extra power would have been turned out; Halkapinar shed was but four miles away. But after ten minutes had passed, no doubts remained; the elusive mocking sound which had haunted me this last half hour was a steam engine fighting its way up the grade and stalling every few yards. I was not the only one to be mesmerized, as droves of children—and a few adults— began to appear at the trackside, presumably to witness what promised to be a remarkable audio visual spectacle. With minutes passing like hours, the tortured sound drew closer; many children ran down the line to find the beleaguered engine; others remained still, their eyes transfixed upon the rocky curve some quarter mile distant.

The familiar face of a German 2-10-2 eventually peeped around the rock. Coughing, slipping and stalling every inch of the way she suddenly gave an animated moan and stopped for a blow-up some fifty yards down the line; she was at the head of a long box freight. Children swarmed around the engine gleefully as the fireman walked towards us carrying a sand bucket from which he liberally dowsed the track. After ten minutes, the crew attempted to get underway; again came tight laboured explosions followed by long blasts as the wheels spun dizzily. Then, without warning, she went into an uncontrollable slip; her wheels were but a blur of movement; as, rocking from side to side, she threw a hideous black mushroom of smoke skywards and sent it drifting menacingly towards Yesildere. The children screamed and cheered, leaping in the air for joy. Why a banker had not been provided I will never know; she had stalled for an hour on this

PLATE 24. *Following pages:* On the opposite side of the Aegean Sea from Izmir lies the Peloponnesus Peninsula of Greece. Here steam engines have already given way to diesels as witness this abandoned metre gauge Class 2 2-6-0T dating back to the 1890s. This moribund study provides a marked contrast with the hustling activities of Izmir.

two-mile climb and another thirty minutes would be needed to clear the summit.

Drawing alongside us, No. 57021—an Izmir engine—stopped again: the track was sanded. Her train stretched way down the bank and though a powerful engine, the design's salient feature was a light axle loading which made slipping an occupational hazard. Thenceforwards, the climb eased slightly and 57021 managed to keep moving at less than walking pace. As the last waggons finally passed, the children pointed excitedly to grooves cut into the rails by 57021's intense slipping. Eventually she cleared the top and disappeared past Sirinyer station— junction for the Buca and Seydikoy routes—situated one mile away. In all she had spent one and a half hours fighting with the gradient and the driver's tenacity must be applauded; many would have given up. But the battle had been won at a cost; the century old Blaenavon rails had been badly lacerated and two passenger trains delayed—although once 57021 cleared Sirinyer, these came up in rapid succession,

One of these passenger trains (G8 No. 44071) goes all out to make up for lost time with the 14.15 to Buca. As she clanked past our cameras, we were thrilled to sense the pungent aroma of locomotive smoke tainted with steam and oil; one of the steam engine's many delightful emanations. Her long conglomeration of ancient rolling stock was full of rustic looking personages, some of whom almost fell out of the windows in their attempt to gaze wide eyed in our direction.

Another dramatic experience was to occur before that afternoon was over. In readiness for the evening trains, I decided to explore a rocky mound; this overlooked the railway and promised to give excellent vantage points for photography. It was impossible to climb from the railway side, so following a stony path to the rear, I looked for a suitable place to climb up. After a while, I encountered an untidy yard filled with wood and bottles obviously belonging to a tradesman. No gate of any description impeded progress and since my only concern was to climb upwards I ventured on and had taken but two paces before a howl went up and an enormous wolf-hound lunged at me, its teeth bared in a vicious snarling roar. By instinct, I flung myself upwards at the rock, gained a foothold that did not exist and managed by inches to avoid its fangs. Meanwhile the dog reached the limit of its twenty-foot chain, jerked to a stop in mid-air and fell back to the ground snarling and barking in a wild frenzy, heaving on its fixture, and mad to get hold of me. Turkish dogs are known to be large, obnoxious and vicious, but this crazed beast surpassed anything I had ever imagined. Such commotion inevitably attracted the children who were soon swarming over

PLATE 25. *Opposite*: French built engines are becoming rare today, but at Izmir in Turkey can be found these 1912 built 2-8-0s from Humboldt of Paris. Delightfully French, she is at the head of a suburban train to Buca.

the rock and in true Turkish style, began to pelt the animal with stones. Against this barrage, the brute fell back immediately and, after being struck by two well aimed missiles, cowered away into its corrugated iron abode, skulking and yelping pathetically. With rocks continuing to clang against the metal, the mangy cur finally slunk from view and was silent.

Though much disturbed by this event, I gained the mound and was in time to catch Humboldt No. 45127 on the evening passenger to Buca.

We had not, as yet, seen a Prussian G10. Only forty-nine are on the T.C.D.D.'s roster and these are widespread throughout Turkey with many working in remote eastern areas; the seven allocated to Izmir being predominantly employed on freight workings. The G10 0-10-0 is another world famous type; they first came to Turkey through the German army during World War One and further examples were built for the Anatolian Railway during the 1920s. However, my aspirations to see one were fulfilled by the 17.40 passenger to Seydikoy, headed by Izmir G10 No. 55044. Watching her toil up the bank, I again faced an identification dilemma; from a distance she looked like a G8, yet the proportions looked too large—was this an illusion created by perspective? For thirty crucial seconds I would have guessed either way but, to my elation, it proved to be a G10—and what potent machines they are! First introduced in Prussia in 1910, they pack their punches on any tough assignment.

Our G10 had left a sooty pall in the valley, as with the sun sinking low, we left the lineside after a truly memorable day. The mosques were calling out for evening prayer, their multi-coloured minarets already illuminated, reminding us that within an hour the vast hillside would be transformed into a tapestry of a million flickering lights. Seven completely different classes had passed through Yesildere that day, German, French and British. Such variety of power and frequency of service, vividly recalled those great lineside days of the past.

9 Steam on the Pilgrim Route to Mecca

After a long, dusty journey from Turkey we finally reached Aleppo, the ancient city of north western Syria. Language posed an almost insuperable problem; all signs were in Arabic and it took over two hours to locate the main station—from which we hoped to obtain directions to the engine sheds. Eagerly we searched the platforms and station yards but no steam engines were to be seen; a water column looked unnervingly parched. Though disappointed, we knew Aleppo to be the main centre for Syrian standard gauge steam with regular workings up to the Turkish border at Cobaney and southwards through the desert to Homs. After further frustrations, compounded by the heat, we obtained directions and eventually reached the roundhouse.

Here there should have been British War Department 2-10-0s; Prussian G8s decked in green with smoke deflectors; Prussian G10s; Borsig Moguls; French 2-8-0s and several nineteenth century 0-6-0s. It was empty. Not one steam locomotive was present; all workings had finished and the engines had mysteriously disappeared.

Our confusion prompted the shedmaster to make several telephone calls and eventually an English speaking fitter arrived. 'You are one week too late,' he said, 'all our steam engines are lying up at Jebrim, seventeen kilometres from here on the Rakka road.' 'Are there none working anywhere?' we enquired. 'No,' he replied, 'their lives are finished.' He offered to take us to the site and so without further delay, we set off; our guide directing us through a labyrinth of long narrow streets leading to the outskirts of Aleppo.

We saw our objective long before reaching it—two depressing rows containing fifty engines. Every Syrian steam engine seemed to be there, including some partly dismantled and several wrecks. One W.D. 2-10-0 was still dripping water from its injectors; these were Syria's most up to date engines, having been frequently employed on the Taurus Express.

As we left the dump, our guide philosophically shrugged his shoulders. 'Too late' he repeated. It was a bitter disappointment relieved only by the thought of narrow gauge antiquities working around Damascus—unless they had finished too! We began our long journey southwards and in vain hope, visited the shed at Homs—a city famous for its huge waterwheels. It was known that G8s and other classes used to work excursion trains from here to Tartus on the Mediterranean Coast but—as expected—the shed was completely empty.

The railways around Damascus offer many fascinations; two routes radiate from a common terminus in the city; one over the mountains to Beirut in the Lebanon, the other—the infamous Hedjaz—heads due south through the desert into Arabia. Both are built to the peculiar Levantine gauge of 3 ft $5\frac{3}{8}$ in.—originally the line was intended to be metre gauge and the deviation was caused by an error in sleeper manufacture.

The line to Beirut was built by a French company under Ottoman concessions in 1895; it crosses the towering Lebanon ranges and, on the Lebanese side, rack worked inclines negotiate gradients of 1 in 14. After the creation of Lebanon in 1919, the line crossed the frontier at Sergayah and today the Syrian section is government operated as the Chemin de Fer Damas–Sergayah (C.D.S.). All services remain steam operated principally by five

PLATE 26. *Above:* The locomotive sheds at Cadem in Damascus offer a fine range of multi-hued veterans. Seen above are a Hartmann 2-8-2 (left), an S.L.M. 2-6-0T and a Jung built 2-6-0T of 1906.

PLATE 27. *Opposite:* The most modern engines on Syria's 1.05 metre gauge lines are these splendid 2-8-2s from Hartmann of Germany. Introduced in 1918, they work the Hedjaz Railway between Damascus and the Jordanian border.

Swiss 2-6-0Ts built by Winterthur in 1894. Initially these worked the non-rack section between Damascus and Rayak but are now solely confined to Syria. Also employed are three Swiss 0-6-2Ts dating from 1894; these were once rack engines but their mechanism is now removed. In 1906, Hartmann of Chemnitz built three 0-4-4-2T Mallets for working up to Rayak; two remain active today, but unfortunately they are only used spasmodically. All three classes were originally coal burners but have since been converted to oil firing.

PLATE 28. *Above:* A study in dereliction at Cadem Works featuring a discarded boiler and a 2-8-0 of early twentieth century design. Both from Hartmann of Chemnitz.

Every male Moslem is required to visit Mahomet's birthplace in Mecca at least once in his lifetime. During the nineteenth century the journey was hazardous in the extreme. The Arabs were reluctantly under Ottoman rule and bands of vulnerable Turkish pilgrims were murdered as they journeyed through the desert. Eventually the Sultan of Turkey authorized the building of a railway between Damascus and Mecca to carry pilgrims in safety; construction began in 1901 and seven laborious years later Medina was reached eight hundred and nine miles to the south where Mahomet is buried. The line (Hedjaz) takes its name from the area alongside the Red Sea in Arabia where the holy city lies.

Trouble dogged every mile, marauding Arabs attacked the workers; the heat was intolerable and violent sandstorms

PLATE 29. *Above:* ... the same scene with heightened atmosphere and night sky.

frequently caused work to be stopped. The Arabs, frantic that their holy city would be defiled, refused to allow the railway past Medina; in wild fervour they invaded the railway construction camp and massacred the work force. The line was destined to go no further and pilgrims had to continue on foot over the remaining two hundred and thirty miles to Mecca.

The railway only carried pilgrims for seven seasons until the outbreak of World War One. The Turks allied with Germany and Arab nationalists, supported by the British under Colonel Lawrence, partly succeeded in driving them from the Hedjaz. In order to prevent enemy reinforcements from getting through, Lawrence blew up large sections of the line and so great was the damage that trains were forced to terminate at Maan in Southern Jordan and today, fifty years later, the southern section remains abandoned despite various attempts to reopen it.

The line now passes through three countries, and all would benefit by its reinstation. In 1963, a consortium of British engineers began work—forty-seven years after the last trains had run. The task was daunting; the section from Maan to Medina is five hundred and twenty-five miles long, and apart from being plagued by the elements—and the damage incurred by Lawrence—thirty miles of embankment had been washed away by violent rains which sweep over the Arabian desert every five

97

PLATE 30. *Opposite:* A
further reflection on the
intense and brooding
atmosphere of the locomotive
graveyard complete with
lamp and hollyhocks. This
wreck also came from
Hartmann during the early
years of this century.

years. Over half the route needed considerable attention and
much track was unserviceable. After considerable work had been
done, the Arab–Israeli war broke out in 1967 and Saudi Arabian
officials ordered work to stop. It was never resumed, and once
again, this ill-fated railway lies abandoned.

The two hundred and ninety-mile section from Damascus to
Maan remains operated as the Hedjaz, but under respective
control by Syria and Jordan, the border being crossed at Deraa
seventy-five miles south of Damascus. The Syrian Hedjaz is
worked by some fascinating steam types and the best way to
appreciate its locomotive history is to visit Cadem Works in
Damascus—a tightly guarded establishment supporting an
enormous coloured headboard over the entrance proclaiming
'Hedjaz Railway'. The works has changed little since being built
and is rife with atmosphere; fragments of engines dating back to
the line's opening are dumped in weed-covered sidings ablaze
with hollyhocks. Both the Hedjaz and Sergayah line engines are
overhauled at Cadem and the works contain—by present-day
standards—a wide variety of classes; and these provided us
with compensation for the tragic cessation of Syria's standard
gauge.

Cadem undertakes major overhauls and entering the erecting
shop, I discovered a Hartmann 2-8-2 stripped to its frame whilst a
sister engine was being reassembled nearby. Pride of the Hedjaz,
these Mikados are employed on the through trains to Amman and
work between Damascus and the Jordanian border. Originally a
dozen worked the line, but today only five remain: six went to
the Palestine railways—and three of these lie abandoned at
Quatrama in Jordan—whilst one was lost to Israel in 1964 when
the Israelis captured a bridge and marooned the engine. Hartmann
built a further ten for the Hedjaz, but after Lawrence wrecked
half the line these were not required; instead they went to the
Dutch East Indies and survive today as the Indonesian State
Railway's D51 Class; they work in northern Java.

On the adjacent road stood an 0-4-4-2T Mallet and nearby was
another Sergayah engine; a Swiss 0-6-2T of 1896. The erecting
shop also contained two lovely looking 2-8-0s from Borsig; only
one other engine of this class remained in Syria and she was
outside in the works' yard awaiting minor attention; fortunately
she would be out working the following day. Alongside stood
No. 91, a smaller and older 2-8-0 from Hartmann dating back to
1906. She was the last survivor of a once prolific class and many
sisters lay forlornly around the works' yard including No. 90—
wrecked when working south along the Hedjaz. Apparently
tanks on military manoeuvres had crossed the track and put it out
of alignment and the train was derailed at speed. In common with

PLATE 31. *Following pages:*
A Syrian 2-6-0T leaves
Sergayah on the Lebanese
border and heads through
the irrigated and fertile
highlands with a passenger
train to Damascus.

99

PLATE 32. *Opposite:* Having worked a passenger train along the Hedjaz Railway from Damascus to Deraa, Hartmann 2-8-0 No 91 prepares to return with freight from Jordan. She is built to the Hedjaz gauge of 1.05 metres.

PLATE 33. *Following pages:* Two Syrian C.D.S. 2-6-0Ts, Swiss built by S.L.M. in 1894 to 1.05 metre gauge, raise steam at Sergayah on the Lebanese border prior to returning to Damascus with passenger trains.

the Borsig, No. 91 only needed detail adjustments and we were to have a thrilling journey with her later that week.

Cadem's rusted hulks contrasted with the working engines but the ultimate discord came from a line of ugly red and yellow diesel units of Hungarian origin which stood lining the works' periphery. These garish machines were awaiting use—allegedly between Damascus and Amman—if agreement could be reached with Jordan upon through operations. Such an arrangement would minimize steam workings; already road and air transportation has greatly reduced traffic on the truncated Hedjaz—failure to rebuild the line having diminished its utility. Immediately alongside the diesel units stood a line of waggons containing rusty boilers lying on their side, a scene visually personifying the transformation of world railways.

Adjoining Cadem Works is the running shed—a modest two road longhouse and focal point for the rich diversity of Syrian steam. As a fleet, Syrian engines must be the most colourful on earth; blue, red, green and black all combine with rusty tones and engines possess such delightful hues as black boilers, rusty smoke boxes and chimneys, red wheels, blue cab and a green tender. Others are predominantly blue—including the Mallets. The shed boasts seven different classes ranging from tank and tender engines to Mallets spanning a building range from 1894 to 1918; not one engine in this multi-coloured fleet being less than sixty years old.

Our researches in Damascus were aided by Hamid El How who, with typical Syrian hospitality, befriended us and insisted we be accommodated at his small farm amid the fertile lands of Dayarra a few miles distant. This green oasis is irrigated by innumerable pumps constantly drawing up water from a natural underground lake formed by winter rains—no rain ever falls between April and October. Arteries of water infiltrate throughout the vast growing lands and Dayarra produces plums, grapes, peaches, apricots and apples along with crops and vegetables which are famous all over Syria.

During our time with Hamid, we lived entirely from the natural foods around; we bathed in the irrigators and rested on the cool farm between ten in the morning and three in the afternoon, as quite apart from the intense heat, the overhead sun made photography impossible. The black desert nights were tranquil except for the chattering pumps all around. Robed figures carrying lanterns flitted through the fields constantly changing the water's direction by building artificial barriers. It was gracious living; during the evenings we sat out in groups, dishes of fruit before us, cigars, and a coffee pot bubbling alongside. Hamid was a lively translator enabling us to speak

PLATE 35. *Above:* After the day's intense sun, the evening's mellow light brings relief to the border country between Syria and Lebanon as an S.L.M. 2-6-0T heads towards Damascus with an evening passenger train.

PLATE 34. *Opposite:* A 1914 built 2-8-0 from Borsig of Berlin trundles the thrice weekly 'mixed' over the lonely branch from Deraa to Bosra on Syria's border with Jordan.

lucidly with his family and friends. After hours of conversation, we slept out under a star-sown sky lulled by warm swirls of desert air—unforgettable experiences from a beautiful country.

The Sergayah line offers many delights; it leaves Damascus through the green Barada River Valley and heads into the hills which separate Syria and Lebanon north of the Golan Heights. In 1976, the Lebanese Civil War prevented through workings to Beirut and all trains terminated at Sergayah. Friday was a special day; a holiday when the inhabitants of Damascus having sweltered all week in the searing heat of this noisy traffic choked city, could get away into the hills to find quietness and cool relief. Two trains ran to serve them, a double header at eight o'clock, followed forty minutes later by a relief with one engine, motive power usually being the Swiss 2-6-0Ts. How the Syrians love those cool hills; picnic spots abound near the small stations and cherries grow in profusion amongst shady groves. On Fridays the trains struggle away from Damascus packed to capacity with people clinging to the sides and even sitting on the roof; songs burst from the coaches and mingle with the engine's barking exhausts. Having left the green valleys, the trains fight their way through a wonderland of yellow rock glistening under a cobalt blue sky and, twisting their way via precipice and tunnel, appear like toys against the looming hillsides; Plate 35.

107

PLATE 36. *Opposite:* One of Syria's C.D.S. 2-6-0Ts raises the pressure at Sergayah having arrived with an excursion train from Damascus.

Having reached Sergayah, all engines turn before taking a siesta in what has now become a quiet terminus. Lines of waggons bound for Beirut stand waiting for the Lebanese tragedy to end. Nearby is a restaurant built alongside a gushing spring and the water, like liquid silver, cascades through the buildings on different levels. Here amid a green sun dappled garden exists a coolness unimaginable in Damascus. A transformation occurs in winter; heavy snowfalls cover the line, adding an extra dimension to its magnificence.

Deraa, in contrast, is situated in open desert country and is an ideal place for watching trains. Apart from the Hedjaz, two other lines converge here; a short branch from Bosra and a line leading out towards Haifa in Israel. Jordanian engines work in from Amman and the yards are shunted by No. 66—a blue 2-6-0T from Jung in 1906 fitted with an old Borsig tender.

Summer mornings at Deraa are often characterized by fog; but once this has lifted the atmosphere remains crystal clear until nightfall. Action begins shortly before 7.00 when an ancient wooden railcar of French origin chunters and rattles its way down the Bosra branch. Meanwhile, Hartmann 2-8-0 No. 91 was being prepared in the shed to work the 8.15 daily mixed to Damascus. The fog usually cleared before this train left and No. 91 made a splendid sight bursting away from Deraa's rock-hewn buildings into the golden desert beyond. A brief lull followed before the Amman–Damascus international express arrived behind a Jordanian Pacific. A Hartmann Mikado had worked down from Damascus with an overnight freight bound for Amman; and the two engines changed trains at Deraa.

What a presence the Jordanian Pacific evoked as she swept into Deraa with her chime whistle blaring impatiently. Black, rugged and utilitarian, her imposing, no-nonsense manner lacked the quaintness associated with Syrian engines. Infinitely more modern than her Syrian counterparts—she was one of five built in Japan by Nippon and delivered to Jordon in 1959. The visitor uncoupled immediately and slipped away to the shed for servicing as a more genteel No. 262 backed on to the train. The express was comprised of six coaches and five vans; gross weight was 188 tons and over four hundred people were on board. At twelve-thirty the Mikado whistled up and, with a full head of steam, eased away from Deraa leaving a multitude of cheering, waving and weeping people on the platform.

PLATE 37. *Following pages:* This 1914 built 2-8-0 from Borsig of Berlin is a fine example of that builder's 'English Phase'. She is seen at Deraa on Syria's Hedjaz Railway.

More activity occurred at the depot and shrouds of jet black oil smoke issued generously across the town. It was Borsig 2-8-0 No. 160; repairs at Cadem were complete and she had returned to Deraa for working the Bosra and Haifa lines. I regarded her as one of the most handsome engines I have ever found; she is

inherently British in appearance and dates back to her maker's 'English Phase'—a period early this century when Borsig consciously produced engines of British styling. At least five of her kind worked the Hedjaz, but a couple were left in Palestine after World War Two. This lovely 2-8-0 was my favourite at Deraa and seemed the epitome of an engine selected to appear in C. Hamilton Ellis' *The Engines that Passed*. She left with the weekly mixed to Bosra—a village world famous for its Roman ruins—and was not scheduled back until early the following morning whereupon she was diagrammed to take a weekly train part-way towards Haifa to collect a train load of tomatoes.

After the Borsig's departure, No. 66, which had stood indolently in the depot all morning, began some heavy shunting sending enormous freight waggons rolling and crashing around the yard and soon a long rake was assembled in readiness for the Jordanian Pacific to take to Amman. Having done its work, the 2-6-0T sidled back to the depot and did not reappear that day. The Pacific made a laboured and noisy departure; her whistle was dual purpose, as threading the suburbs, she used a clear deep-toned call. Having watched the Amman freight leave, we were invited to join the shed foreman for tea and in broken French, were able to communicate something of our enthusiasm for Syria's steam engines. Shortly before seven o'clock a shrill whistle rang out from the Damascus direction; it was the mixed returning with No. 91 and clouds of dust followed her as she clanked in with seven vehicles. She did her own shunting—presumably to avoid disturbing the ever somnolent No. 66—and, after half an hour, retired to shed. Thenceforth a quietness fell over Deraa; the day's workings were complete.

The following day arrangements had been made for us to ride with No. 91 on the 'mixed' to Damascus and at eight o'clock we joined our engine which stood gleaming in the morning sunlight with steam pouring high into the air from its safety valves. A final round of oil was applied to the motion and cotton waste was handed out before a ceremonial ringing from the huge station bell heralded our departure. The Syrian authorities are extremely concerned about the track condition on the Hedjaz; no trains must exceed 55 km. per hour and all engines are fitted with a speed recording graph. As we began the seventy-five mile journey, No. 91 bucked wildly on the old trackbed, giving an elated impression of speed. Semi-desert lay all around us, partly cultivated but almost completely devoid of buildings or animals. Maximum speed was maintained for mile after mile, but an absence of landmarks belied our progress. Sixteen stops were scheduled; eight at stations and the remainder at isolated desert halts for farmers and land workers travelling on business.

112

Cold tea was passed around the footplate and as our journey progressed, this became increasingly appreciated; for once the sun was up the footplate's blinding heat was inseparable from the temperature outside. Our driver told us proudly that he had two wives—not a particular extravagance—his brother had three! The conviviality of a locomotive footplate is well known by all railway lovers and our crew were thrilled that we had come from Britain to ride the Hedjaz.

The sprawling suburbs of Damascus appeared all too soon and No. 91's whistle screached continuously for innumerable crossings; the thronged streets were jammed with cars, their horns blaring for each yard of laboured access. We passed Cadem Shed and Works and caught tantalizing glimpses of blue Mallets, Mikados and long-funnelled Swiss tanks before we joined the Sergayah line and drew to a stop against the buffers at mid-day, having covered the seventy-five miles in three and three-quarter hours, a start-to-stop average of exactly twenty miles per hour.

We had three hours in which to service No. 91 before returning to Deraa. Noon in mid-summer is the worst possible time to be in Damascus but, necessity being the mother of invention, it is Syrian practice to sell fresh fruit drinks, consisting of crushed oranges and shredded ice, on every corner. Our 2-8-0 had gone to Cadem for refuelling and its train was marshalled by No. 61—a Jung 2-6-0T. On the return journey we were to carry three old tenders full of water; these would be put off at intermittent stations for the local populace. At three o'clock No. 91 got its heavy train underway and barked out of Damascus. A few miles out, the crew pointed to the spot where No. 90—working this very train—derailed on the track damaged by tanks.

Our return trip was unforgettable; the engine's lovely pounding roar, the aroma of steam and oil and more lashings of cold tea. Outside a golden sunlight—almost mystical in its brilliance and clarity—illuminated the yellow desert under a deep blue sky. At each station women carrying buckets or cut-down oil drums crowded round the engine for hot water, begging in animated Arabic for the injector exhaust pipe to be put on. Our crew were happy to comply, and even entrusted me to the task, but the queues were too long and, had we supplied everyone, our journey would have taken twice as long. Hospitality followed us everywhere, and one station master, who knew little English, begged us to be his guests for one night and continue our journey with the next day's train.

The sun sank rapidly as we steamed southwards and I watched the desert hues turn from yellow to gold and through to a lustrous bronze before the radiance lost its purity and combined with black to form a cool twilight. Shortly after seven o'clock, the

STEAM ON THE PILGRIM ROUTE TO MECCA

Haifa line swung in on our right and the lights of Deraa appeared on the horizon; we were home after a one hundred and fifty mile trip and eight hours on the footplate.

'Alors vous n'avez pas dormi', called the shed foreman from the platform. 'Non monsieur,' I replied, 'vos locomotives sont trop passionnantes'. After a little shunting No. 91 retired to the shed and, half an hour later, looked like a dead engine; such is the rapid transformation of personality when oil burning engines take their rest.

PLATE 38. *Opposite:* C.F.H. Hartmann 2-8-0 No. 91 traverses the Syrian section of the Hedjaz with the daily 'mixed' from Deraa to Damascus.

115

10 A Cottage by the Shed

It is almost thirty years since the evening I cycled over the fields with a friend to a village south of Leicester on the Midland main line. We were under ten years of age. Reaching the railway bridge, my friend suggested we stop and wait for a train and, leaning our bikes against the brickwork, we stood on our crossbars and commanded a fine view of the line looking towards Leicester. Standing there, I felt greatly intrigued to see a train pass, though I knew nothing about engines. After ten minutes, a coal train came into sight; the engine was working hard making a superb roaring sound and pumping grey smoke into the air.

I can clearly recall it to be a Stanier 8F—so vivid was the experience. Mesmerized, I watched the locomotive approach until its heavy roar became too frightening and we leapt down from our crossbars. The smoke struck the bridge's underside and puthered up in a mighty cloud, providing me with my first experience of the ultimate in exotic perfume. The bridge shook alarmingly as heavy waggons piled high with shiny black coal rolled beneath on their way to London. Another bridge was situated a quarter of a mile further along the line, and as we watched the 8F forging its way towards it, we wondered if the train would stretch between the two. To our joy, the engine reached the distant bridge as the last waggon and guard's van passed beneath ours and the entire cutting was filled with a symmetrical coal train echoing the line's curvature.

This incident represented a turning point in my life and I was back at the bridge on several successive evenings, until I became so immersed in the pleasures offered by a steam main line, that much of my remaining childhood was to be spent in that place. Stanier 8Fs were the predominant type amongst many classes seen, but in those post-war days, locomotive ABCs were not easy to obtain and I used to put down the engine numbers in large spidery writing in a dog-eared exercise book with bright red covers (I describe it clearly, as I remember it with such great

PLATE 39. *Opposite:* A brace of American built 'Skyliner' 2-10-0s avidly boil up at Irmak prior to working heavy trains over the mountainous route to Zonguldak on Turkey's Black Sea Coast.

affection). When I obtained an ABC, there began the ceremony of transferring the numbers from my notebook—but I digress into another theme.

Today, I live in the village next to that bridge and, from my study, can see the curved stonework against which we stood. The line remains busy, but with a soulless uniformity compared with those golden years gone by and the really heavy freights—once so ably hauled by the 8Fs—are seldom seen nowadays. But memories remain; physical reminders—a Midland Railway trespass notice—forty shillings penalty; blackened brickwork; or even a piece of bull-head track in Kilby Siding. More beautiful is the spiritual memory; many is the night, long after our village is asleep, when I hear the phantom call of a Stanier 8 tackling the grade southwards from Kilby Bridge with a Toton–Brent coal haul. So real are the phantoms, I could almost ask the people living in our village square—'did you hear that heavy coal train last night; did it wake you?'

The first 8F was completed in summer 1935 at Crewe, as a standard design intended to supplement the L.M.S.'s many 0-8-0s. When war broke out in 1939, over a hundred had been built and having proved themselves highly capable, the 8Fs were selected by the Ministry of Supply as a war-time engine and a further 208 were built by the North British and Beyer Peacock for W.D. service. Orders were also given for 8Fs to be used all over Britain to cope with heavy war-time loads and the type was subsequently built at Horwich, Doncaster, Darlington, Ashford, Eastleigh, Brighton and Swindon.

The W.D. took some fifty 8Fs from L.M.S. stock and thus had over two hundred and fifty under their control. However, the German advance prevented them from being used in Europe, but in 1941, an urgent need arose for locomotives in the Middle East and the W.D. shipped the majority of their 8Fs for use in Egypt, Palestine, Turkey and Iran. After the war, many were destined to remain abroad, although thirty-nine were returned to Crewe in 1947; these were incorporated into L.M.S. stock after overhaul. A further five came back in 1952, this time to Derby Works and I clearly remember seeing them standing in a line complete with Arabic numerals and what we took (as young boys) to be bullet holes. These engines created much interest and after repairs went to Longmoor. In 1957, three passed to B.R. stock as Nos. 48773/5; the other two remained in W.D. service until withdrawn in 1959.

Turkey retained the twenty sent in 1942 and they became T.C.D.D. Nos. 45151–45170. Well over one hundred went to Iran, but many were later redistributed, including twelve purchased by the Iraqi State Railways. However, twenty-two remained working in Iran until the 1960s. Italy received fifteen; these

118

worked along the Adriatic line as FS Class 737 until withdrawn in the late 1950s; others remained active in Israel until 1958. Fourteen were broken up for spares in Egypt during 1944, whilst others were named after members of the Royal Engineers who had obtained the Victoria Cross. Egypt's last working examples were scrapped by the State Railways in 1963.

Six hundred and sixty-nine eventually passed into B.R. stock and all were transferred to the former L.M.S. system—taking the running numbers 48000–48775. By 1953, they were working from forty-eight different sheds with particular concentrations at Willesden, Toton and Wellingborough. Even in 1964, their ranks were largely intact and some survived until the end of British steam in 1967 when they were confined to working from a few depots in Lancashire.

Fortunately, the 8F is not extinct in Britain, one being employed on the Severn Valley Railway. Carrying her old L.M.S. number 8233, she is ex-B.R. No. 48773—one of the five returned to Derby from the Middle East in 1952. This engine was originally built for the W.D. by North British in 1940 as their No. 24607 and, along with many sisters, was briefly loaned to the L.M.S. before shipment to Egypt in 1941. In addition most Turkish 8Fs remain active today and some are believed to be still at work in Iraq.

Ten years after their demise on B.R., I next saw a 'Stanier 8' at Irmak—a remote junction east of the Turkish capital. Irmak is approached by a dirt road which curves round a hillside and, from afar, I saw the 8F busily shunting; you can imagine how thrilling it was to renew acquaintance with this childhood type in a remote Turkish hamlet. I was amused by her number—45161—as being that of a related Stanier 'Black 5' in Britain. Sixty children swarmed into the depot yard behind us; by appearance they might have come from the film set of 'Oliver'. Fortuitously, these were banished by a kindly looking man who bade us to follow him to his cottage, which stood on a small bank overlooking the shed yard. Here we met his wife and baby daughter. It was impossible to communicate, other than by basic sign language, but once Cemalettin, our host, understood our purpose he was emphatic that we should stay in his cottage over the days we had planned photography around Irmak. Cemalettin was a shedman whose job it was to service the engines which came on shed.

We were amongst people unknown to us, with whom little communication could be made; their country was totally alien, yet we felt completely at home. Possibly, the friendly presence of engines—especially the 8F—had something to do with our contentment; one often dreams that such idyllic situations will present themselves amongst the general rigours of touring.

Having made acquaintances, Cemalettin accompanied us to watch No. 45161 shunt the yard. Today all Turkey's 8Fs are confined to shunting and general pilot duties. On her right hand side, No. 45161 was completely English, but on the other Westinghouse pumps and cylinders gave a somewhat foreign aspect. The spark arrestor clipped around her chimney lay idly to one side and was barely noticeable; but the saddest change was the substitution of Stanier's deep-toned whistle for the shrill high-pitched affair applied to Turkish engines in general. Indeed whenever our proud 8F whistled up during yard shunts, she sounded more like an ex-Great Eastern J15 0-6-0!

Otherwise, No. 45161 was identical with her sisters which passed beneath my boyhood bridge. She even possessed those layers of sooty grime which revealed neither colour nor rust—a characteristic Willesden grey! Mounting her cab the layout was instantly familiar; what fine workmanship had been bestowed; the 8F's longevity was hardly surprising. Watching her, I thought how well this Turkish exile would look trundling a long loose-coupled goods over the old L.N.W. main line.

Irmak—a two-road longhouse—is predominantly concerned with workings up to Zonguldak; the main line from Istanbul to eastern Turkey being diesel operated. 'Middle East' 2-8-2s work the passenger trains and Skyliner 2-10-0s the freight. One other class was evident—an ex-Prussian G82 which shuffled around with a supercilious air on shed pilot duty. On our first evening, six Skyliners were present and a Middle East 2-8-2 stood ready for the overnight passenger to Zonguldak.

Later that night, the 8F began some heavy shunting. Her exhaust beats were so familiar, so uniquely Stanier 8F, that I realized how individualistic an engine's sound really is, although many sound identical superficially. When running through the yard with a full train she might have been out on a main line. Even the trace of slipping on a damp rail after Turkey's light summer rains, conjured up vivid recollections of those lineside days on the drag from Kilby to Wistow. Towards the end of each shunt, she would cut off and free-wheel the last few yards with a musical roll and plonking from her wheels and valve gear. How these beautiful sounds carried me back down the years; back to the soft Leicestershire countryside and childhood. This 'ghost of Crewe' was a glowing fragment of history: to feel homesick in that enchanted place was impossible.

Breakfast the following morning was at six; the table was laid on a terrace completely enclosed in slatted wicker fencing entwined with grapes; though dense, the foliage permitted

PLATE 41. *Following pages:* The halcyon days of North American railroading are eloquently recalled by these two T.C.D.D. 'Skyliners' pulling out of Irmak with an enormous rake of empties for Zonguldak. Gathering storm clouds complete the drama.

mottles of golden sunlight to pour through. Engines hissed outside and by peering through broken slats one could see what trains were passing or what engines were coming on shed. Some holes only provided a partial view of an engine—like the 'bits and pieces' puzzles which appeared in the old *Trains Illustrated*. During meal times these holes were my link with outside, but if anything particularly interesting occurred, I would leave the table, walk down the garden, lean over a small fence, and bask openly in the sooty haze which enveloped the cottage and lightly dusted whatever food was put out at table.

After breakfast, we went to the lineside in readiness for a 'Middle East' arriving with the Zonguldak–Ankara passenger. She was running two hours behind time—a minimal delay in a country where trains are frequently ten hours late. Our wait proved worthwhile; speeding towards us through the rock cuttings came spotless 'Middle East' No. 46201, embellished by red wheels and enormous smoke deflectors. She looked unlike a war engine; a racy, low-slung design capable of good speed and overshadowed by a high tender piled with coal. The 2-8-2 stood majestically in Irmak station with nine coaches until, after a brief stop, she dashed away towards Ankara. Later that morning, No. 46201 returned for servicing, but we had little time to admire this distinguished visitor as she was diagrammed to return with the overnight passenger after which dirty and unblinkered No. 46204 would be the class's sole representative at Irmak. Two hundred 'Middle Easts' were built in 1942 for Britain's Ministry of Supply by Baldwin, Alco and Lima and were shipped direct to Suez for work in Syria, Palestine and Persia. Fifty-three remain in Turkey as T.C.D.D. Nos. 46201–46253, the last fifteen being obtained from Iran in 1955.

Two Skyliners were being prepared for respective freights. Having already marshalled their trains, the 8F took one of its periodical breaks from shunting and, instead of retiring to the shed, or placidly simmering by a water column—as is normally the case with yard shunters—her habit was to become utterly lifeless and stand sandwiched between waggons. In this mode of indolence she was difficult to locate and looked—if ever an engine can—truly 'fed up'.

About four o'clock, Cemalettin came looking for us to report that two Skyliners were preparing to leave with a freight for Karabük. Hastily we passed to the far end of the yard just as the two engines were backing on to a seemingly endless freight. It had been intended to run the trains separately but a last minute decision combined them.

Skyliners running in tandem are rare and with over an hour's sunlight left, we decided to take chase. A dirt road paralleled the

Zonguldak line for some miles and, along with Cemalettin, we set off at breakneck speed. Half a mile out, we crossed the track a hundred yards ahead of the two giants which bore upon us with whistles blowing, cylinder cocks open and fires fully made up. Meandering haycarts did little to impede our progress, but the rough surface and blind corners made fast driving impossible. It soon became apparent that the two engine drivers—who knew Cemalettin—were entering into the spirit of the chase and with wide throttles the American giants began to gain ground. The road deteriorated badly and our vehicle was bouncing like a cork in water, forcing us to slow down. Gradually we suffered the humiliation of being overtaken and, after four miles we were running neck and neck. It was a tremendous thrill to watch these Skyliners at speed; their small wheels spinning dizzily—I remembered the high speed runs enacted by B.R.'s 9F 2-10-0s. The goggled crews waved enthusiastically as they drew ahead. Dashing rolls of black smoke across a rocky landscape, the Skyliners disappeared from view leaving only racing waggons at our side. It was futile to continue the chase; dejectedly we returned to Irmak.

A marvellous meal had been prepared for us and many of Cemalettin's relatives had come to the cottage to meet his English visitors. They were amazed to hear about our interest in the 'Churchills'—as the Turks call their 8Fs. How could we explain the class's historical significance? And that years ago identical engines ran through our village in England.

The 'Middle East' left on time and at ten o'clock Cemalettin went to begin night duty, leaving us sitting on the terrace listening to No. 45161 renew its endeavours. A train from Istanbul–Kars on the Russian border came in headed by a diesel; the seventeen illuminated coaches stretching through the station and halfway along the yard. It was calling at remote places in eastern Turkey; places where steam abounded, but a hostile reception would be given to a foreigner.

About midnight, I went to join Cemalettin and found him raking out a Skyliner's firegrate. Sheets of crimson cinders rained into the pits; it was hot work for a summer evening and a daunting task if one considered the three other engines waiting their turn on the pits. I remained for two hours, enjoying the incomparable atmosphere engendered by steam giants being serviced, until the heat and excitement induced its inevitable drowsiness, and I returned to the cottage and went to bed. There is something comforting and reassuring about the proximity of a steam railway; far from keeping one awake, it promotes restful slumber and, for the second evening in succession, I was lulled by the 8F's music.

Shortly before leaving Britain for Turkey, I lectured to a learned society in the Midlands and afterwards, a retired engine driver stood talking with me. He had spent many years on the L.M.S. and we recalled some notable steam classes. When I mentioned that Stanier 8s remained at work in Turkey and Iraq he blandly replied: 'Why not? They'll stand it.' Such poetic simplicity cannot easily be surpassed.

PLATE 42. *Opposite:* An American built Middle East 2-8-2 heads away from Irmak with a light freight for Cankiri.

11 Karabük Steel Works

The mountainous route from Irmak to Zonguldak is greatly revered today; it is the haunt of 'Skyliner' 2-10-0s. How superbly these Pennsylvanian built Decapods blend with the rugged terrain. Steam railways acquire an extra dimension of magnificence amid such wild grandeur; tunnels, viaducts and miles of circuitous slogging gradients through barren hills add much to their appeal. Enormous hauls are undertaken; traffic from the Black Sea port of Zonguldak combines with heavy tonnages to and from Ulku Steel Works at Karabük — not least of which is iron ore from Sivas in the east. Accordingly, these mighty engines are worked to full capacity. Banking occurs at various points throughout the 414 km. slog; until recently this was performed by ex-U.S. Army S160s, but sadly most of these are now dumped in the reserve at Behicbey and Skyliners, made redundant by dieselization, are more frequently used today. Banking is undertaken in both directions from Sumucak — a lonely crossing point which provides an ideal stopping place on the journey from Irmak to Karabük.

South of the station, the line crosses a curved river bridge before entering a tunnel and I was fortunate to catch an immaculate Skyliner heading a trainload of coal bound for the interior from collieries along the Black Sea coast. Notice the green vegetation around the river bed, against a backdrop of harsh boulder strewn semi-steppe country.

Towards evening, one of the banking engines began to 'boil up'; she moved off the ash pits, took coal and water and with safety valves singing softly, stood ready to bank a northbound train. We assumed a position on a hillside overlooking a curve situated one mile north of the station, and long before the train reached us, we heard the syncopated stacatto bark of the two engines as they fought away from Sumucak; the curves and rock cuttings causing their exhausts to be ever changing in volume and texture. As the train approached, we noticed the macabre

contents of the leading waggons; they contained the last remains
of a standard German 44 class 3-cylinder 2-10-0 which, having
been cut up at Sivas Works, was in conveyance to Karabük for
melting down. Turkey received forty-eight of these superb
engines from France in 1955, when they were rendered
superfluous by the Thionville to Valenciennes electrification.
They worked the Taurus route from Adana to Konya, but by
1976, most had succumbed to dieselization and gone to Behicbey
dump for storage. The stricken engine's number—56742—was
chalked over the remains.

Approaching Karabük, one encounters the unmistakeable
smell of steel making long before the town and its industry comes
into view. Upon sighting the works, I was reminded of Corby in
Northamptonshire. Karabük has grown up around steel making
and many of the town's inhabitants are employed by the plant.
The Turkish hero, Kemal Ataturk, was instrumental in planning
the works and his death in 1938 coincided with the opening. It was
the country's first steelworks and its remote situation surrounded by an amphitheatre of hills was chosen as security
against air attacks: similarly, Ataturk selected Ankara as the new
Turkish capital, its secure inland position providing a safeguard
against advancing enemies.

Karabük's ore comes from hundreds of miles away in the east,
but is seventy per cent. iron compared with Corby's locally
produced thirty per cent. Much uncertainty existed about
Karabük's steam fleet; but it was known to be varied with a
predomination of British types.

Reaching the works' periphery, we peered into the gloomy
structures. The industrial clamour was laced by plumes of pure
white steam from a coking plant; behind appeared drifts of
orange smoke as charges of iron ore entered the furnaces. Steam
and smoke emanated from a dozen points and flanged piping like
an endless snake, embraced every structure giving the works a
surrealistic appearance. High above the turmoil, a bleeder pipe
spouted a shimmering tongue of flame with the darting rapidity
of a serpent's fang; this was the crowning sceptre in a shrine
devoted to a technological god. 'Look, over there,' I pointed
enthusiastically as a chunky eight wheeled side tank decked in
green and typically British surreptitiously plodded into sight,
weaving an unknown path amid the structures. This, we were to
discover, was one of two built by Bagnall's of Stafford in 1937 for
the works opening. A sheet of dark red flame shot from the coke
ovens and the distant hills became momentarily obscured by
smoke; a line of waggons could be seen moving against this fiery
outrage and there appeared a familiar green saddle tank. It was a
standard Hawthorn Leslie, a classic British type whose outlines

PLATE 43. *Following pages:*
Some fine action as a Vulcan
(Wilkes Barre) 'Skyliner'
heads southwards through
the mountains with a coal
train from the Black Sea
collieries. Eighty-eight of
these huge 2-10-0s were built
for Turkey between 1947/9.

have changed little since the beginning of the twentieth century.

Spurred on by these tantalizing glimpses, we endeavoured to gain entry; but, after innumerable telephone calls made on our behalf—which accumulatively occupied several hours—we were obliged to wait until morning for a final answer. Thus, with frustration tempered by optimism, we repaired to a hotel for the night. After years of touring, it becomes second nature to select hotels which command a 'railway view' and about eleven-thirty that night, I heard a freight 'whistle up' and prepare to leave. I left the room and went to a corridor window which overlooked the line. A Skyliner was approaching and as I waited for it to pass a ghastly yellow glow lit up the town. It was as if the moon had suddenly become an enormous light bulb—everything was illuminated—the river flowing through town; distant houses perched around the valley; and even the barren hills in all their rocky detail way beyond Karabük. Having reached a fearful peak, the glow faded and—like a black shadow—darkness closed in again, first extinguishing the distant hilltops and sweeping inwards until only the immediate surrounds were affected by a lurid incandescence more akin to a moving shadow than actual light. The entire effect had taken less than a minute. No coke oven or blast furnace could produce such a crescendo of brilliance; yet the action must have come from the steel works. My deliberations were interrupted by a reappearance of the phenomena; even brighter this time as, once again, the darkness was progressively chased outwards to reveal distant objects in minute detail, until having fully illuminated Karabük, the barren rocky hills beyond again became as bright as day. Transfixed, I watched this incredible event occur a further three times until the sombre orange shadows died away completely and only the vague pin-points of city lights indicated that Karabük existed at all.

The following morning we arrived at the works to find that negotiations on our behalf had achieved an 'open sesame' and, in a company jeep, we were driven to the Transport Manager's office—conveniently situated alongside the blast furnaces. Having been cordially received, a tray of Turkish tea and oaten biscuits was produced.

Although no English was spoken, it soon became obvious that our presence created a dilemma; no one seemed certain whether we could make pictures or even be shown around. As the delay was likely to be lengthy, it was fortuitous that our office overlooked a busy junction with trains passing from all parts of the works.

Two Hawthorn Leslie tanks were visible, one taking water, the other slowly reversing a rake of crimson ingots fresh from the steel furnace. These engines were identical with their British relations, except for a Turkish 'Star and Crescent' affixed in brass to their cabsides. A grimy 0-6-0T of German origin flitted by 'light engine'; returning minutes later pulling hopper waggons filled with coke. Next, came a more modern German engine built by Jung, but she quickly disappeared from view leaving the track clear for a Bagnall 0-8-0T to arrive with a line of empty slag ladles; she drew past slowly and placed the rake neatly under a blast furnace in readiness for refilling. Minutes later, a prolonged piping whistle heralded another 'through train' and a Schwartzkopf built 0-8-0T puffed by with a string of waggons comprised of platelayer's flats and scrap metal. Two hours passed watching these exciting operations; the British and German engines making a fascinating contrast.

The restrictions applicable to 'foreigners' in key industrial establishments did not extend to the motive power record book, and this was made freely available for our inspection. Every Karabük engine was listed in detail. Six Hawthorn Leslie type 0-6-0ST were shown, numbered 3301–6; all except one were active—number 3304 allegedly having been scrapped five years previously. Numbers 3301/2 were Hawthorn Leslie engines built in 1937 and delivered the following year. The other four came from Robert Stephenson and Hawthorn (the amalgam of Hawthorn Leslie and Robert Stephenson). Two of these were delivered in war-time, arriving in January 1944; the others followed in 1948. All were identical except larger coal bunkers were fitted by R.S.H. Our observations, combined with Karabük's records, enabled us to enumerate the fourteen engines of five different classes on the company's roster, but their assertion that number 3304 had been scrapped was gloriously dispelled when she appeared in the yard outside—not only alive and well—but heading an applaudable trainload of coke. The completed roster appears at the end of this chapter.

A further hour's waiting produced more delightful glimpses of operations around the yard interspersed by further tea, this time with huge Turkish chocolates. Eventually, authority was given for us to make one picture of a locomotive and we chose the Bagnall 0-8-0T standing nearby with slag ladles. At last we were amongst the engines. Seeing No. 4401 close up reminded me of a 1937-built Bagnall 4-8-2T I had found at a Transvaal gold mine in South Africa.* These two classes were probably built side by side—their works numbers running consecutively—and were almost certainly designed by the same man.

A steady trickle of golden waste poured into the ladles as No. 4401 stood patiently simmering. When filled she was to carry

PLATE 44. *Following pages:* The ghastly glare of molten waste from steel making illuminates the town of Karabük in Northern Turkey. Undertaking the ceremony during the early hours is a standard gauge 0-8-0T built by Bagnalls of Stafford in 1937.

* *Steam Safari*, 'Last Steam Locomotives of the World' series.

133

them through the plant to a slagbank whereupon the cauldrons are tipped to disgorge their contents down an incline. Suddenly I realized what had caused that traumatic glare the previous evening. I imagined these enormous cauldrons spewing out rivers of molten waste. No wonder Karabük became aglow: the mystery was solved.

Our enjoyment of this remarkable place was shattered when a workman dashed up to us shouting and waving his arms; another man blew a whistle. A wave of panic swept the yard as we were pushed, at a run, back towards the office. Initially, I thought objections were being made to our photography, until the Transport manager pointed in horror to large globules of liquid waste seeping ominously from the ladle's base. A sizeable flow had developed and the men were terrified in case the ladle opened up completely and deposited a sea of liquid across the yard. In a flurry of steam, the Bagnall charged forward and cautiously backed up to the offending waggon: the fireman coupled up with a shunting pole and, with whistle blaring continuously, No. 4401 raced away towards the tip. The ladle swayed behind depositing a stream of liquid gold in its wake.

Behind the furnaces lay enormous mounds of iron ore and an S160 2-8-0 was shunting empty hoppers in the T.C.D.D. connection: Karabük has a steel making capacity of $\frac{3}{4}$ million tons a year. Nearby, lay tangled piles of scrap twenty feet high awaiting charge to the furnace. Every possible metal commodity was apparent from old baths to battleships including many steam engine remains. More fragments of Reichsbahn 3 cylinder 2-10-0s were recognizable; along with some older Mogul and Tank classes; positive identification was difficult, the pieces having been much reduced to facilitate the rail journey from Sivas.

The slagbank was a thrilling experience and is possibly unique in using steam locomotives. It consists of a high spur of land whose edges fall away sharply into a valley to form the banks which in their steely grey barrenness resemble the moon's surface. A single track line leads from the works, but upon reaching the 'peninsula' this diverges to enable simultaneous tipping from either side. Two tips occur on each side per shift making a total of a dozen drops in twenty-four hours. Up to seven ladles are carried on one train and, having placed them strategically above the bank, the engine moves back twenty yards and the cauldrons are tipped by a mechanically operated pulley. The cauldrons are filled and tipped in the same sequence and sometimes the first one will have set like a huge blancmange. It will part reluctantly in one moulded piece, slither down the bank and suddenly smash into a thousand pieces; only at the very centre will there be liquid. Each ladle gains in excitement

PLATE 45. *Previous pages:* One of Karabük Steel Works' Bagnall 0-8-0Ts of 1937 patiently waits for the ladles to fill with molten steel.

until the final one is reached; this will invariably be molten and will fall in a gushing river; a Bagnall 0-8-0T is depicted performing the duty.

Spectacular as this is, the night drops are frightening and unforgettable: the searing heat can be felt two hundred yards away, whilst the towering hills above Karabük glow ominously and momentarily appear like huge tidal waves poised to crash down and engulf the entire valley. Imagine the whirr of ropes; the creaking of an upturned cauldron; the cataclysmic blaze of light and the hideous crackle as the waste explodes upon hitting the ground. From a molten river, terrifying fingers pour into every nook and cranny until the cauldron is empty: by this time the dislodged holocaust swirls and writhes impotently along the bottom of the bank.

STEAM ROSTER, KARABÜK STEEL WORKS

Engine No.	Wheel Arrangement	Builder	Works No.	Date
3301	0-6-0 ST	Hawthorn Leslie	3921	1937
3302	,,	,,	3922	,,
3303	,,	Robert Stephenson and Hawthorn	7082	1943
3304	,,	,,	7083	,,
3305	,,	,,	7417	1948
3306	,,	,,	7418	,,
3307	0-6-0 T	Jung	11567	1952
3308	,,	Henschel	19372	1922
3309	,,	,,	19374	,,
3310	0-6-0 WT	,,	19380	,,
4401	0-8-0 T	Bagnall	2578	1937
4402	,,		2579	,,
4403	,,	Schwartzkopf	9202	1928
4404	,,	,,	9201	1928

12 Along the Black Sea Coast

PLATE 46. *Opposite:* A T.C.D.D. 'Skyliner' 2-10-0 heads northwards through Sumucak on the Zonguldak line. Behind the leading coach can be seen the cut up remains of German type 44 three-cylinder 2-10-0s en route from Sivas to Karabük Steel Works.

A broken fence separated the depot yard from rough meadow land where sheep grazed and two small boys flew kites. Sitting in a little thatched patio, I looked towards Eregli and watched clouds of smoke and vapour rise from the distant steelworks. The sun had already fallen below the surrounding hills, lighting only the very highest points and causing them to shine like glowing beacons. Wealds of orange and crimson light rippled across the horizon of the Black Sea; just visible one mile distant. A little kettle began to boil furiously and threatened to spurt its contents across the floor, until a shed man swept it from the flame and dutifully mashed a pot of the inimitable Turkish tea. It was a perfect summer evening; nothing stirred in the balmy atmosphere except the shed cat optimistically scouring the ashpits. Two tall figures dressed in suits and trilby hats appeared far across the meadow; they approached with casual deliberation, the stroll being as important as their destination. Reaching the yard, they stepped around the huge Prussian $G8^2$—which stood alongside the patio—and joined our circle. What better for these off-duty engine-men to walk to the shed of an evening for a quiet cup of tea, a smoke and a chat with their fellows? Rugged in manner and aspect, they possessed a seasoned maturity; I was reminded that Turkey was once the centre of the vast Ottoman Empire.

The aroma of tea and tobacco merged with oily sulphurous emanations from our $G8^2$, as she stood gurgling softly outside. The workmen's apparel reminded me of bygone days at home; I might have been at an English sub-shed during the 1930s, days when sitting out on summer evenings with neighbours and workmates was part of the social order. Another $G8^2$ stood in the shed entrance and a banging shovel, followed by the rumble of falling coal, indicated that she was being prepared for work. Later that evening, both engines were booked off shed.

Eregli is a sub-shed in the truest sense; only three $G8^2$s being

PLATE 47. *Following pages:* The T.C.D.D. operate several famous ex-Prussian designs as witness this pair of G82 2-8-0s at Eregli. These two examples work coal hauls from Armutcuk Collieries to Eregli Docks—a system isolated from the main T.C.D.D. network.

141

allocated for working to Armutcuk Colliery some seventeen kilometres along the Black Sea coast. The line is totally divorced from the remainder of the T.C.D.D. network and the engines are brought from Zonguldak by ship. However, a well equipped workshop at Eregli depot ensures that locomotives do not have to be changed very frequently. Armutcuk's coal is either shipped from Eregli Docks or goes to the vast steel complex to which this modern town owes its existence. Some Turkish maps—by naivety or wishful thinking—show Eregli and Zonguldak connected by a coastal line; certainly Eregli shed is over commodious for three engines and its long administrative corridors are little used, but whether the link will eventually be built is anyone's guess.

The line to Armutcuk is especially scenic; it follows the sea and 1 in 40 gradients occur along the coastal alignment. Armutcuk is a remote mining village surrounded on all sides by hills; these act as a windshield and the area is permanently characterized by a heavy smell of coal. Having barked their way up from Eregli, the G8²s pass through the colliery yard and deposit the empty waggons on a gentle incline from which they roll by gravity to a coal grading plant for filling. When full, the waggons are collected by an electrically driven cable; the engine attaches to the rake and, running tender first, proceeds to Eregli: no shunting operations being necessary.

The G8²s belong to the celebrated family of Prussian goods engines. Quite apart from providing much of Europe with highly standardized classes, Prussian design set a basis for the 'Reichsbahn Standards' prepared from 1920 onwards for a unified Germany. From the turn of the century, Prussia concentrated upon building standard classes with large numbers of engines and many still survive today amid Europe's last steam locomotives. The spreading of Prussian types occurred either by German war action, by territorial changes, or as a result of Germany's vigorous export market. Several Prussian classes rank amongst the all-time greats in steam history whilst collectively they form an intrinsic part of the German school of design.

During the 1890s, Prussia's G7 series included both 0-8-0s and 2-8-0s, but in 1912 came the G8 with superheater, piston-valves and Walschaerts valve gear. World War One greatly swelled the ranks of this superb goods engine and some five thousand three hundred were eventually built to see use in twenty countries and three continents. About the same time, an engine of similar power, but lighter axle loading, was needed; and so were born the G10 Class 0-10-0s—shown on Plate 23. Apart from being highly significant within themselves, the G10's boiler was interchangeable with the Prussian P8 4-6-0, introduced in 1906.

PLATE 48. *Opposite:* As twilight descends at Catalagzi, a Bagnall 0-6-0PT storms out of the yard with a rake of empties. Notice the mixed gauge track at this Turkish colliery.

PLATE 49. *Following pages:* A smoky study showing a different aspect of the Bagnall 0-6-0 Pannier Tanks at work on the colliery system at Catalagzi in Turkey.

The P8s also had a widespread distribution and are regarded today as the forerunner of the modern mixed-traffic 4-6-0. Some still survive in Eastern Europe, especially Poland—large parts of which were once in Prussia. Both the G10 and P8 exceeded a total of three thousand five hundred engines.

In 1915, Henschel of Kassel prepared designs for an advancement on the G10, and in 1917, they produced the first G12. This was a mighty three-cylinder 2-10-0 with a capacity to haul one thousand ton trains over 1 in 200 gradients at 25 m.p.h. The G12 represented a different phase in Prussian design by having a Belpair firebox and plate-frames. Over one thousand five hundred were built and they became the immediate forerunner of the D.R.'s famous 44 class three-cylinder 2-10-0s built after 1938 to some one thousand seven hundred and fifty engines. The 44s disappeared from West Germany in 1976, but others survive in the eastern sector. In 1919, the G12's character passed into the powerful two-cylinder $G8^2$ 2-8-0. More than one thousand were built, including sixty-two for Turkey between 1927/35. These, which came from Nohab and Tubize, were numbered 4500–62 by the T.C.D.D. It was fascinating to see the $G8^2$'s family delineation at Eregli; they looked like a reduced D.R. 44 2-10-0 and from certain angles might have been mistaken for one; so akin are the two classes to their G12 ancestor. Notice the $G8^2$'s highly characteristic chimney.

An interesting story occurs concerning a lack of Teutonic precision in the $G8^2$ design. The 2-8-0 was in effect a reduced G12 with one pair of coupled wheels removed and the boiler, firebox and tubes shortened accordingly. But unfortunately, the same diameter tube was retained for the shorter boiler causing the gases to enter the smokebox at too high a temperature and Richard Wagner—the Reichsbahn's motive-power chief— claimed the engines suffered greatly as a result! Despite this, the $G8^2$ has given some spirited performances and certainly handled their duties masterfully at Eregli.

The Eregli to Zonguldak road is not for the faint hearted; its rocky unmade surface is narrow and strewn with enormous pot holes. As it twists around the hillsides, a bottomless chasm opens up on one side threatening to swallow up the inattentive driver. Zonguldak is an important coal producing centre and terminus of the four hundred and fourteen kilometre line from Irmak. A Bagnall 0-6-0T shunts in the docks, but most steam activity is situated at Catalagzi nine kilometres away. Here, one finds the huge engine sheds, and next to them, a splendid mixed gauge colliery worked by steam.

The shed has a fascinating allocation: American built 'Middle East' type 2-8-2s for working suburban trains to Caycuma and long distance passenger to Irmak; 'Skyliners' for main line and coal hauls; and Prussian G8s for yard shunting. Several S160 2-8-0s are also present along with a couple of old German built Moguls, dating back to the Anatolian and Baghdad Railways. The blackened long house is steeped in atmosphere; huge pulsating engines roar amid the gloom and pieces of locomotive anatomy lie everywhere. Heaps of clinker appear at random.

Adjoining Catalagzi goods yard is the E.K.I. Colliery with its Bagnall 0-6-0 Pannier Tanks operating on metre gauge lines. Four were built by Bagnall's in 1942; all remain active today. The collieries' standard gauge lines are worked by an ex-T.C.D.D. No. 3305; one of ten supplied by Henschel in 1918 for the Turkish army during World War One. On her right is another Pannier Tank.

This colliery is one of Turkey's finest railway sights. Twice a day, a workmen's passenger train leaves the yards for the distant mines. The train is always double headed by Pannier tanks and consists of nine covered goods waggons improvised as coaches; archaic in the extreme, they are four wheelers fitted with bench-seats but no windows. Upon departure, the two Bagnalls thunder across the yard with full regulators and screaming whistles in an effort to work up sufficient impetus for the climb to the colliery, their conglomerated rolling stock rattling and swaying behind. What a delightful contrast this rustic train makes with the double-headed 'Skyliners' which sweep past on the main line with ultra-heavy coal trains.

Some Catalagzi G8s are in a chaotic state of decrepitude. Caked in soot and grime, one engine had a large pile of soot permanently adorning the smokebox top. Her piston extension casings were missing, leaving the rods to flail about in a highly dangerous way. Large portions of boiler sheeting had rotted away partially revealing the inner firebox, and the cylinder cover was missing from one side. When involved in a particularly heavy shunt, she leaked badly and disappeared immediately in a thick ball of steam; wheezing like an anguished baby, she rivalled the old L.N.W.R. 0-8-0s in B.R. days. Other G8s were in better condition, including one revolutionized in looks by a handsome pair of smoke deflectors.

Whatever their inadequacies, the decrepit G8s were vastly superior to the epic produced by the Catalagzi Electricity Works. They used the ultimate proliferation in shunting engines—a truncated steam crane built by Rodley of Leeds. Minus her jib, she shunts the waggons by rope-haulage! After the coals have been delivered—usually by a 'Skyliner'—this apparition appears on the adjacent track and one by one draws the waggons to a small incline: the rope is swiftly disconnected and they roll

PLATE 50. *Following pages:* A fine contrast in the colliery exchange sidings at Catalagzi near Zonguldak. A T.C.D.D. standard gauge 0-6-0T from Henschel 1918 (*left*) boils up alongside a British metre gauge Pannier Tank from Bagnalls of Stafford in 1942.

gently to the station's coal bunkers. Shunting with a gay abandon, she emits a rapidly repeated 'wush hush' sound; smoke and fire puther from her chimney and the flywheel spins neurotically.

The industry around Zonguldak is of considerable national importance to Turkey. On Catalagzi's grimy streets, industrial workers mix with those engaged in social customs of a bygone age. Men and youth predominate; and are the sole occupants of the shabby tea bars—unlike neighbouring Zonguldak, no women are allowed within. Donkey carts vie with coal lorries as this backward village with its timeless customs confronts the technological world.

The surrounding hills partially melt into a blue haze, residue from the electricity station's belching chimneys. Even as far away as Zonguldak, one can look out across the Black Sea on a clear and tranquil evening, and witness ominous dark streaks across the horizon: a permanent signpost to Catalagzi with its omnipresent industry and symphony of falling coal and steam whistles.

East of Catalagzi, the contrast is difficult to reconcile; the line passes through a scenic wonderland: on one side is the non-tidal Black Sea in all its shimmering radiance, on the other, a wide panoramic vista of sweeping green hills. Having traversed deep gorges between hill and cliff, the trains plunge into long tunnels and emerge alongside isolated beaches radiant in their timeless beauty and devoid of man and his endeavours.

This section vividly recalls the Great Western main line around Dawlish, but instead of Castles it is the preserve of 'Skyliners'. These classes could hardly be less similar yet both are thoroughbreds; the Castles represent a pinnacle in Great Western design, the Skyliners a pinnacle in latter day American practice. Freight traffic over these coastal reaches is supplemented by 'Middle East' 2-8-2s, dashing along with suburban passenger trains to Caycuma. Turkey's steam fleet—rather like Yugoslavia's—is cosmopolitan, with well-known classes from Britain, Germany and America. Turkey only built two steam locomotives, modern 2-10-0s based on a Skoda design. Named 'Black Wolf' and 'Grey Wolf', they were built in 1961 at Eskisehir and Sivas respectively.

At Hisarönü, the line finally turns inland and heads through hill country for Karabük. Hisarönü is characterized by a large brickworks, and the raw clay is brought by ship to a long jetty extending into the sea. This jetty carries a branch from the main line and the clay is transferred by crane into rail waggons for its final half-mile journey to the factory. The duty is performed by different classes ranging from Henschel 0-6-0Ts; or one of the splendid 34001 series of Moguls.

PLATE 51. *Previous pages:* One of the handsome Moguls built by Borsig in 1911 for the old Baghdad Railway wanders on to the pier at Hisarönü on Turkey's Black Sea coast.

She was built by Borsig in 1911; one of thirty period Moguls built in Germany for the Anatolian and Baghdad Railways between 1911/14. Their $15\frac{1}{2}$ ton axle loading was too high for certain sections and some were fitted with a small pair of carrying wheels situated between the middle and rear driving axles. This rare innovation reduced the loading to $13\frac{1}{2}$ tons. During World War One, the British captured four of these Moguls en route from Germany to Turkey, and diverted them to work in Egypt. In later years, some passed to the Iraqi Railways but the majority went to the T.C.D.D. Several survive, but all additional axles have been removed.

I was fortunate to discover this Mogul at Hisarönü, as the previous week's clay turns had been handled by an S160. Work is governed by the frequency of ships, but in between duties, the jetty engine sometimes helps heavy main line goods trains on the sharp climb southwards.

As we have seen, the Black Sea coast around Zonguldak has a fine tapestry of classes to interest the visitor; 0-8-0, 2-8-0. 2-10-0, 2-8-2, 2-6-0, 0-6-0T and 0-6-0PTs—covering nine different types built in three countries embracing two gauges. Excellent variety for the late 1970s and proof that the great steam age is not yet past. The availability of locally mined coal means the section between Zonguldak and Karabük will be one of the last to retain steam traction in Turkey. But the winds of change are ever blowing, and soon this now world famous area will fade back into obscurity under the ever encroaching diesel.

PLATE 52. *Opposite*:
Dismembered pieces of
locomotives often combine to
produce an enhanced feeling
of reality. This rare study
was made at Cadem Works,
Damascus.

Index

	Page	Plate No.
JZ class 53 2–8–2T		plate on p. 6
Andrew Barclay 0–4–0ST	12, 16–20	**1, 2, 4**
Hunslet Austerity 0–6–0ST	23–32	**5, 6**
W. D. Austerity 2–10–0	33–40, 74, 93	**8**
Stanier 8F 2–8–0	33, 39, 117–27	**40**
Hunslet Austerity 8F 2–8–0	33, 34, 36, 38, 39	
Major Marsh S160	33, 34, 40, 75, 128, 132, 149, 155	
Reichsbahn class 52 'Kreigs-lokomotive' 2–10–0	36, 82	**9**
Brunel 'Firefly' class 2–2–2	41	
G. W. 4–4–0 'City of Truro'	44	
G. W. 5006 'Treganna Castle'	44	
L.N.E.R. A4 'Silver Link'	44	
A4 Pacific 'Mallard'	44	
Reichsbahn 01 Pacific	46–53	**10, 11**
East German 2–10–2T	50, 56, 60	**12**
Meyer 0–4–4–0T	54, 56	
German Mallet 0–4–4–0T	55, 96, 99	**13, 15**
German 2–6–2T	56	**14**
Henschel 0–6–0T	56, 139, 149, 155	**50**
'Feldbahn' 0–8–0T	60	
Floridsdorf 0–6–0 Fireless	60–8	**16, 17**
Andrew Barclay Fireless	68	
JZ class 170 2–8–0	70	
JZ class 180, 80, 80·9, 0–10–0	70	
Kg class 0–10–0	72	
Sudbahn class 580 2–10–0	72, 74	**18**
U.S.A.T.C. 0–6–0T	74, 85	**19**
Breda & Ansaldo 2–10–2	75	
Ottoman Railway 2–8–2	79	
Prussian G8 0–8–0	79, 86, 90, 144, 149	**20**
Stephenson 2–8–2	82	
Corpet Louvet 2–10–0	82, 85, 86	**22**

Humboldt 2–8–0	82, 86, 92, 109	25
German 2–10–2	86, 87, 90	21
Prussion G10 0–10–0	92, 144, 148	23
Class 2 2–6–0T		24
S.L.M. 2–6–0T	96, 107, 109	26, 31, 33, 35, 36
Hartmann 2–8–2 'Mikado'	99, 109	26, 27, 31
Borsig 2–8–0	99, 109–12	34
Jordanian Pacific	109, 112	
Jung 2–6–0T	109, 113	26
Hartmann 2–8–0	109, 112, 113	28, 32, 38
'Skyliner' 2–10–0	121–25, 128, 149	39, 41, 43, 46
'Middle East' 2–8–2	121, 124, 149, 154	42
Prussian G82 class 2–8–0	121, 141–2, 148	frontispiece, 47
German 44 class 2–10–0	129, 148	
Bagnall 0–8–0T	131, 133, 139	44, 45
Hawthorn Leslie 0–6–0ST	132, 133, 139	
Bagnall 4–8–2T	133	
Schwartzkopf 0–8–0T	133, 139	
Stephenson & Hawthorn 0–6–0ST	133, 139	3, 79
Jung 0–6–0T	133, 139	
Henschel 0–6–0WT	139	
Prussian P8 4–6–0	148	
Prussian G12 2–10–0	148	
Bagnall 0–6–0PT	149	48, 49, 50
Borsig Mogul	149, 155	51
Turkish 2–10–0	154	

Colin Garratt

Colin Garratt lives in a small Leicestershire village overlooking the bridge from which he watched his first steam train in 1949. That bridge was the starting-point of his world-wide travels recording in words and on film the survivors of the steam age. He has already covered some thirty countries and is still running a desperate race against time, as every country in the world has declared against steam traction and many have already eradicated it.

Colin Garratt undertakes an annual season of lectures for the Pentacon Praktica manufacturers of cameras and lenses, and his personal lectures and slide tape portfolios – based on his travels – are popular with schools and societies. He also takes photographic classes and contributes to various magazines.

Details of his lectures are available from Monica Gladdle, 'Carlestrough Cottage', Shangton Road, Tur Langton, Leicestershire. Tel: East Langton (08 5884) 438.